Dedicated to my Beloved Sister Rahat Iqbal

&

And to my Best Friend Ali Raza Yaseen

“We don’t want to conquer the cosmos, we simply want to extend the boundaries of Earth to the frontiers of the cosmos... We don’t know what to do with other worlds. We don’t need other worlds. We need a mirror.”

- Solaris by Stanisław Lem (1961)

Contents

Preface

When I was a child, curiosity was my closest friend and often my only one. I wanted to understand how everything worked. Why did the rain fall? Why did plants grow toward the sun? Why did stars shine? Every natural process caught my attention, begging for answers. Science became my escape. For any kid growing up in a broken family, you need something like that.... a way to rise above the pain at home, something that takes you away, even if just for a little while.

In those quiet moments, hiding in books or gazing at the sky, my imagination grew strong. It learned to create worlds when the real one felt too small or too difficult. That habit never left me.

During my Biotechnology graduation, my curiosity only expanded. Labs revealed the tiny miracles inside cells - how life repairs itself, how it adapts. Then I began reading about quantum physics. My imagination jumped to a new level. I started seeing humans defying gods, building settlements on distant planets, rewriting the very rules of existence. The boundary between science and myth blurred. I noticed people playing with fire - not the fire of stories, but the fire of knowledge, creation, and power that could save us or destroy us.

To give shape to these wild ideas, I explored further. I read history to understand how humans have always reached too far.

I delved into religion to see what people believe about creation and gods. I studied quantum biology - the strange ways quantum effects might influence life, like in photosynthesis or how birds navigate. I read biotechnology books and research papers on stem cells, genetics, and the limits of what's possible. Piece by piece, I gathered knowledge like puzzle pieces.

All of this led to one decision: I had to express these thoughts in a book. A book that blends real science with big questions. A story about humanity's endless desire to know more and to become more, even when it hurts.

I named it **The Last Prometheus** because of an old Greek myth. Prometheus was the Titan who stole fire from the gods and gave it to humans. He gave us knowledge, technology, the spark to build and dream - but he paid a terrible price. Chained to a rock, his liver was eaten daily, only to regrow. He became a symbol of rebellion, of bringing light to the world at great cost.

Today, we are all somewhat like Prometheus. We steal fire from nature - through biotechnology, quantum insights, and genetic editing - and use it to change life itself. We create, we heal, we dream of immortality.

If you've ever felt small, ever wondered what humans could become, ever feared or hoped for the future we are shaping - thank you for opening this book. Let's walk this path together.

Bilal Iqbal
Faisalabad, Pakistan
February 2026

I

The Stolen Fire (Myth, History, & The God of the Gaps)

Part I lays the foundation by linking ancient myths to modern science. Prometheus and Norse sagas are re-framed as early metaphors for humanity's drive to shape biology. As "Unwitting Alchemists," we began biotech with wolf domestication and grain fermentation (Biotech 0.1). The "God of the Gaps" filled unknowns - until we replaced fate's threads with DNA's double helix.

1

The Spark & The Clay

"Myth is the secret opening through which the inexhaustible energies of the cosmos pour into human cultural manifestation." — *Joseph Campbell*

The sky above the world was dark, with a multitude of stars shining. Below this sky, a human was sitting by a fire that was out at the entrance of a cave. The human was just sitting there looking up at the sky and the stars with the cave behind him and the dying fire in front of him, and the ancient world around him. He looked at his hands. His hands were really rough and dirty. They had a lot of scars on them. His hands had helped him make his life better. He used his hands to make knives out of stones so he could hunt for food. He used his hands to sew skins of animals to make clothes that would keep him warm. He even learned how to make fire with his hands so his family would be safe when it was dark. His hands had done all these things for him and his family. That was a good thing. He had changed his world with his hands. As he looked at his hands moving, he felt really frustrated. The world around him was something he could change, but his own body was different. He could not change the biology of his body. Humans can make things like a spear but they cannot make their own bodies strong and invincible. They can build a shelter to leave or to hide but they cannot stop their body from hurting or aging. A person can be really good at using tools he creates but he can't control the working of the tools in his own body. Their own body can hold them back sometimes. But at that time, something became clear: humans are born with a desire to do things, but their bodies are not able to keep up with what they want to do. Humans have a spark inside them that wants to achieve things, but the limitations of physical bodies cannot keep up with this spark.

This cave was the laboratory. It was not because of the things it had inside. Because it was the first place where a living creature dared to dream, dare to knock on the doors of imagination. As Einstein said, "Imagination is more important than knowledge. For knowledge is limited to all we now know and understand, while imagination embraces the entire world, stimulating progress, giving birth to evolution." The cave was a place where the creature could think about being more than what nature intended for the creature.

We want to be able to break the rules of nature. This is where our oldest stories come from. Our stories have had myths of people who lived forever or could heal any wound for thousands of years. We told these stories for a reason. They were not just to entertain people. We told these stories because they showed what we wanted the most: to have control over our bodies, the bodies that are made of something like clay. We want to have power over the clay we are made of.

Think about the famous stories of creation. You have the story of Prometheus, who made people from clay. Then there are the Norse stories where life was made from wood. What is really interesting is that these stories all have something in common. The stories of creation tell us that life is something that someone makes and shapes, that creation is something that is made by someone or something. The stories of creation are really saying that life is created and crafted, that it does not just happen on its own. In these stories, the clay is like our bodies, our bones, our blood, and our skin. The spark is what makes us alive and smart; it is what makes us human beings. For a long time, people believed that only something bigger than us, like the heavens, fate, or gods, could give the spark to the clay. We thought that people were, like, finished statues and that even when they were all done, they could not be changed at all.

Today, things are different for us. We've started to see that the clay is not just a finished statue sculptured by something divine, but it is something that is still being worked on. The clay is a living thing that is still taking shape. This is not about replacing God or disproving myths. It is about the fact that, finally, humans have reached a point where our tools are starting to live up to our imagination. We have always wanted to take control of the power of life and health, just as I say that we want to steal the fire as Prometheus did. Humanity wants the power of life and health in its hands. We are moving from being people who just watch their lives happen to being people who control them. In the coming chapters, we will discuss how far we've come in controlling the creation of our own species and the others. The powers that were once imagined in the form of gods are now in our own hands. We have stolen the fire from the gods. We are the species that has finally learned how to pick up the tools and finish the work that nature started.

Prometheus: The First Bio-Hacker

In Greece, the pattern of thoughts in people like us did not just show up out of thin air. The old stories say that a powerful being named Prometheus was in charge of making the species. He sat down by the river in a place called Panopeus and started doing something that sounds a lot like what scientists do in a laboratory today: he combined things to create something from imagination. He combined dirt with water to make humans. The Greek myths really tell us that Prometheus was the one who made people. He was like a creator making humans, from the earth and water. The story of Prometheus and the humans is an interesting one. When you think about it, Prometheus was not a person who made sculptures. He was actually working with the things that make up living things. What is clay and water really, if not a way of describing the building blocks of life and the water that is needed to make them work? Our bodies are made up of a lot of water and complicated chains of units. Prometheus was the person to figured out that the earth itself had the potential to be something really amazing if its parts were put together in the right way. Prometheus was working with the ingredients of life, like biopolymers and hydration, and he understood that these things could be used to create something new. The idea that our bodies are made up of water and complex molecules is really important. Prometheus seemed to know that the dust of the earth could be used to make something great if it was organized correctly.

There was a problem, though. These early humans were really weak. They were like clay that just would not hold its shape. The early humans would shiver in the dark. They were at the mercy of every predator and every disease that came their way. The early humans had the form of humans but lacked the power to survive on their own. So Prometheus wanted to save the people he created. He did something rebellious to do this. Prometheus took fire from the top of Mount Olympus. This was a big deal because Mount Olympus is where the gods live. Prometheus stole fire to help the people he created. He got this fire from the home of the gods, on Mount Olympus and gave it to humans.

Let us look closely at what "fire" actually represents here. In their language, this was called Pyr - πῦρ. Some people believe that the word for fire is Technè. Techne means skill or craft, not fire. Prometheus gave humans fire (pyr), which then allowed them to learn techne: arts and techniques. So in that sense, πῦρ can be considered as Technè. Anyway, it was not a burning branch; it was the knowledge of transformation, the knowledge that fire gives us. Fire is the tool that allows us to change one thing into another thing. It turns grain into bread, cold ore into iron, and darkness into light. Prometheus gave us fire that changed everything. Fire is like the starting point for technology. It allowed us to do things that other animals cannot do. We did not have to wait for our bodies to change over time to be able to survive. Humans could make their own things to keep warm and to protect themselves. Prometheus and fire, and technology, helped us to do this. We could use technology to make our lives better and easier. Other animals have to wait for nature to help them. Humans can use technology to make their own way. This was the bio-hack. The guy Prometheus figured out that our biology was not enough for us humans to really do well. We needed to be able to change the world around us. Later on, we needed to be able to change our own bodies, the human body, and our biology to make our biology better.

The Gods were really upset about this theft. They got very angry with Prometheus for giving people the power to change into things. So Zeus punished Prometheus by tying him to a cold rock in the Caucasus Mountains. Every day, a huge eagle would come down and rip out his liver. Then every night his liver would grow back, I mean, regenerate. The next day, the eagle would come again and eat his liver all over again. The Gods did this to Prometheus because he gave people the power of transformation.

For a long time, people have seen this as a story about people suffering a lot. When I look at this story, I see something that really surprises me because I work with living things and technology. The liver is a special part of our body that can fix itself in a way that no other part can. If someone takes a piece of a liver out, it will be back to normal in just a few weeks. This is very different from other important parts like the heart or the lungs, or the brain. The liver is the one that can heal itself completely. The liver has a power that lets it grow back to its normal size. No other part inside us can do this as the liver can. This should make you wonder: did the ancient storytellers know something about biology that we are only just proving in labs today? They must have had some idea about the body. How did the ancient storytellers pick the liver, which is actually the one organ that behaves like a self-healing machine and not the heart or intestine or stomach?

The story of Prometheus might not have been about Gods

being angry. It could have been the ancients trying to tell us something about how our bodies can fix themselves. They noticed that life can repair its damage, like it can regrow its own parts. The people then saw that humans have a special power inside them, a kind of immortal spark inside our bodies, and they made a story about it. The myth of Prometheus is really interesting because it talks about the idea of regeneration. The myth of Prometheus says that we have a hidden ability to heal ourselves, and that is a cool thing to think about. We used to be like people who made things from clay and water by the river. Now we are working with DNA and proteins in a laboratory. This is a change for us. In the heart of every innovator, Prometheus breathes still, stealing the divine fire of healing, the wisdom to mend our climate, and the power to reanimate lost life. The ultimate authority over our evolution is no longer with the gods; it is being stolen back by us, by The Last Prometheus.

We are finally understanding that the "Fire" wasn't just for cooking or warmth. It was a hint. It was an invitation to look at our own bodies - our regenerating livers, our beating hearts, our complex brains - and realize that we are not finished products. We are a work in progress, and the tools of transformation are finally in our hands.

The Golem: The Secret of the Living Word

If the story of Prometheus is about the physical materials of life - the clay and the fire - then the legend of the Golem is about something even more modern: The Code.

The famous version of this story comes from 16th-century Prague. This story is about a man named Rabbi Loew. Rabbi

Loew wanted to create a protector for his people. He did not ask God to do it for him. Rabbi Loew went down to the banks of the river. Shaped a massive figure, out of mud and clay, just like Prometheus did. When the figure was finished, it was a lifeless statue. The statue of mud and clay did not have a spark of life. The Rabbi did not use fire to wake it up; he used language instead. That is what he did: he wrote a secret word on a piece of paper and put it in the Golem's mouth. In some stories, he cut out the word Emet, which means Truth, in Hebrew and put it on the Golem's forehead. When the letters touched the clay, the Golem opened its eyes, stood up and breathed. The Golem did what it was told to do.

The Golem was not made alive by a spell like we think it was. It was made alive by a specific set of information. In the story of the Golem, if you changed one letter of the special word, the Golem would stop working or fall apart into dust. The Golem had to have the word right, or it would not work. The information that made the Golem alive had to be perfect.

This sounds really familiar, right? I mean, when you look at the DNA, you see millions of letter combinations. If you cut one letter of a specific part of DNA, the molecule loses its ability to work properly. DNA is not something else; it is a code that has four letters, which are A, C, G, and T. This code is part of the chemistry of our cells. Think of it like a story where a Rabbi puts words into a Golem's mouth to control it. Today, scientists are doing something like this. They are putting sequences of DNA into bacteria, plants and even human cells to tell them what to do. When we use something called CRISPR to change a gene, it is like changing the words on the Golem's forehead. We are changing the DNA code that tells our cells what is true and how to behave. The DNA code is like a set of instructions that says what our cells should do. When we change it, we are changing the instructions. This is a thing. We are changing the DNA code, which is the code of life, and this can have big effects on our cells and our bodies. We are rewriting the instructions to fix a mistake or add an ability. This is because the old instructions were not correct, or they did not have something that we need now. So we are changing the instructions for the instructions to be better. The new instructions will have the fix for the mistake or the new ability that we want.

For centuries, the legend of the Golem was regarded as either a miracle or a fairy tale. However, by shifting our perspective to view the Golem through a scientific lens, we can analyze the story as a laboratory experiment. If we treat this ancient narrative as a technical case study rather than a myth, what insights could we uncover about its construction and function?

- **The Clay**: This is our body's hardware. It is made up of proteins, lipids and carbon. These are the things that make up our bodies, like The Clay.
- **The Word**: This is the Software. The Software has all the information that tells the hardware what it needs to do. The Software is like a set of instructions for the hardware. As we have a set of instructions in our cells.

The Golem is teaching us that life is not just made up of useless things; it is made up of things that have a purpose. If you have a bunch of carbon and water, it is a puddle of water. If you have a bunch of carbon and water that are arranged in a special way, then you have a human being. The Golem is showing us that life is like a code, and when things are arranged according to this code, they become something more than just a pile of stuff. The Golem is talking about life. It is made up of things that are arranged specially, like a human being is made up of carbon and water that are arranged according to a specific code.

The legend also has a warning that makes sense for our world today. In a lot of versions of the story, the Golem gets so powerful that the humans who made the Golem start to lose control of him. The Golem grows fast and becomes too strong, and eventually the Rabbi has to remove the word to turn the Golem back into dust. This is why we should be careful with our technology. We are learning how life works. We are learning how to tell cells to fight cancer. We are making plants grow in the desert. We are even changing how humans are born. For the first time, we are in control. We are the ones making the rules for life. But what happens when this technology gets too strong? Will it turn against us? Can we turn it into dust, or will it turn us into dirt? This is a very big deal. This is the dangerous game of biotechnology. Anyhow, this Golem thing tells us that life is not something that only the Gods know about and that we can never figure out. Life is like a puzzle that we can solve. If we know the steps to take and we understand what is going on, then we can make something happen. We can take something that's not alive, like clay, and make it come to life. Life is like a puzzle, and if we have the right information, we can solve it and make things work.

We used to think that only God could know the plan of life. When we started to figure out the human genome, we found out that the answer was not something from above. It was actually inside our cells the whole time. We have found the code. We have read the human genome letters. The human genome is like a book, and we have finally been able to read it. The human genome is an important discovery. We are still learning about the genome and what it can tell us about life. But one thing is for sure: the human genome is the key to understanding how our bodies work. The human genome is a map that can help us navigate the complexities of life. We are lucky to have found the genome and to be able to read its letters.

The legend of the Golem teaches us that the only difference between inanimate mud and a living person is information. For a long time, we thought the "code of life" was a sacred mystery that humans should never touch. That has changed. We have finally figured out how life works, and we've realized that biology isn't a divine secret - it is a set of instructions we can control. We are no longer just observers; we are the creators. The real challenge is no longer about whether we are allowed to change ourselves. Instead, it's about whether we are brave enough to erase our human weaknesses and design something that lives forever.

The Death of the Potter

For a long time, people have looked at the world around them and seen how amazing it is. They have seen the complicated patterns in the eye, the way our skin can fix itself when it gets hurt, and the strange thing that is our consciousness. People have looked at all these things and thought to themselves: there has to be someone who made all this. If you see a pot, you know someone made it on a wheel. If you see something that was designed, you know someone did the designing. The idea of the world and everything in it being an accident just does not seem right. People have always believed that there is a creator, a maker, a person who designed everything, like a

Potter who makes pots. This Potter was really important to us. He helped us with the things we did not understand. The Potter lived in the areas where our knowledge was not complete. As time went on, something big started to happen slowly. We did not just look at the pot and say it is nice; we started to think about the clay it was made from. We began to wonder about the clay that made the pot. That is when things started to change. The Potter and the pot and the clay were all.

When I look through a microscope as a biotechnologist, I see things in a way. We are finding out that the amazing things people used to believe in are actually things our bodies do. The idea of the "breath of life" is really just our bodies using glucose for energy. The "threads of fate" people talk about are actually the connections between the parts of our DNA. The "clay" is not a lump of dirt; it is a tiny machine that is always moving and follows the rules of physics and chemistry.

The Death of the Potter does not say that a creator cannot exist. It is just that the Potter is not needed anymore for the process. We found the plans that were hidden under the workbench. We figured out that we can shape the clay ourselves with our hands. The Death of the Potter is a change. We do not need the Potter to make things happen. The Death of the Potter means we can do it on our own.

As we leave the world of stories and legends behind and enter the world of science and experiments, we come across some ideas that people use to support the notion of a perfect plan created by a higher power. The idea of a fixed divine design is something that people really believe in. To move forward and write the part of human history, we need to take a close look at these big ideas with a clear and truthful mindset, like a scientist who does not take anything for granted. We must consider the idea of a fixed design and really think about what it means.

The Argument of Irreducible Complexity (The "Watchmaker" Argument)

Some people think that life is just too complicated to be altered or improved. They claim that if you take away or modify one thing in a cell, the entire system falls apart. So they believe that life must have been perfectly designed from the beginning. They say that life is too complex to be changed or hacked. Life, they argue, is something that must have been carefully planned and designed, with every single part working perfectly. The idea that life can be hacked or improved is not something they agree with. They think that life as it is must have been perfectly designed from the start.

This argument is based on the idea that the current design of life is the only possible design. Biotechnology shows us that this is not true. When we use CRISPR to remove a gene or add one, we are not breaking the thing that was made; we are making it better. This shows us that biology is like a set of building blocks that can be changed and improved. Life is not something that will fall apart easily; it is a flexible set of instructions. If the person who created us, the "Potter," really did make us, then they made us in a way that allows us to make improvements like an "Open Source" system that can be modified and updated. The biotechnology of CRISPR proves that the design of life is not fixed; it can be made better. My opinion is that life is an adaptable code, and the "Potter" made us with the ability to improve and upgrade ourselves. We are not a finished statue; we are a beta version.

The Argument of the "Sacred Blueprint."

People often say that we should not mess with biotechnology because of the idea of the "Sacred Blueprint". The "Sacred Blueprint" is the idea that the human genome is something special that should not be changed. The human genome is like a set of instructions. Some people think these instructions are perfect and should never be altered. They think of DNA as something that was written a long time ago and should not be changed. I am a biotechnologist, and when I look at the human genome, I do not see something that is perfect and finished. The human genome is like something that is still being worked on. It is a little messy, beautiful, and chaotic.

If the genome were like a plan made by a perfect creator, we would think it should be simple and work really well. But what we actually see is that the genome has a lot of mistakes in it, like things that happened by accident. The genome is full of these accidents, which is really surprising if we think of it as a "Sacred Blueprint" made by a perfect Potter. The genome has a history of accidents that have built up over time.

Our bodies have a lot of viruses in them. 8% of our DNA is made up of these ancient viruses that got into our ancestors a very long time ago. We have the scars of these old infections in every single cell of our bodies. Our DNA is also full of things called jumping genes and repetitive sequences. The jumping genes are also known as transposons. These things often do not do much of anything except make noise in our DNA. The ancient viruses and the jumping genes are a part of what makes up our genome. If this is supposed to be a plan from a power, why does it have so much junk in it, like the digital version of unwanted emails? The digital world is full of this kind of thing, really annoying. I mean, if this is a plan, it should be better; it should not have all this spam in it. A divine plan should be nice and clean, not filled with digital junk.

The truth is that the genome is something like a diary that our ancestors kept. This diary has all the things that happened to them by chance and all the bad things they managed to avoid over a very long time. The genome was not made to be perfect. It was made to be good enough to keep our ancestors alive enough to have babies. The genome is like a story of how our ancestors survived. That is what it is. Nature does not care if you tend to get Alzheimer's disease when you are seventy years old. This is because, from the perspective of nature, you have already done what you were supposed to do by the time you're thirty years old. Nature thinks you have already served your purpose by then. Alzheimer's disease is not a concern for nature, at this point, because you have already grown up and had children by the time you're thirty.

But we are editing the "Unchangeable". When people criticize us for using CRISPR-Cas9, they say we are playing God. They think that the way things happen in nature is the best. Let us look at what really happens with the so-called Sacred Blueprint when doctors are treating patients in a hospital.

Imagine a child who is born with Tay-Sachs disease or Huntington's. The genetic information that this child has, which is like a plan for their body, has a mistake in it. This mistake is one wrong letter, a single fucking letter out of more than three billion letters. For a scientist who does not believe in God, a mistake like this is not sacred; it is a problem that causes a lot of suffering for the child. The idea of a "Sacred Blueprint" does not make sense when you think about a child who has to suffer because of a mistake in their genetic code, like a typo that you would find in a book. Tay-Sachs disease and Huntington's are very serious, and they are caused by mistakes in the information, so you can see why a scientist might not think that these mistakes are sacred.

When we use a scalpel to change a Cytosine to a Thymine, we are not crossing a line that we should not cross. We are doing what people have always done: we are fixing something that is broken. If we have the technology to stop a heart from failing or a brain from getting sick, then the idea of a Cytosine and a Thymine and the whole molecular scalpel thing is not a rule. It is an idea. The molecular scalpel and the Cytosine, and the Thymine are tools that we can use to make people better. As the Last Prometheus, we think that we can come up with something that is really good. The Last Prometheus is what we call ourselves. We believe the Last Prometheus can do a lot better. The fear of playing God usually comes from the idea that we're not smart enough to handle the fire. People ask: What if we make a mistake and change something with the fire that we did not mean to change with the fire? The fire is a concern for people when they think about playing God with the fire. This is a question about something called off-target effects, but it is not about religion. It is a problem that engineers need to solve. We do not stop making airplanes just because the first ones had accidents. Instead, we figure out how to make engines. We do the thing with editing DNA. We do not stop editing DNA just because it is hard to do. We keep working to get it right.

As human beings, our morality isn't based on obeying an ancient blueprint. It is based on reducing suffering and expanding potential. If the God or Gods left us with a genome full of errors, then it is our moral duty to pick up the pen and start editing. We are transitioning from being a species that is written to a species that writes.

The Argument of Human "Essence."

The argument is that people are worried about changing their bodies. They think that if we live forever, we will not be human anymore. Many people believe that because we can die and we are not perfect, that is what makes us human. They think that our mortality and the things that are wrong with us are what make us the way we are. The idea of living or stealing eternity is scary to them because they think it will take away the things that make us human.

I think this is a romantic way to look at suffering. It is said that death gives your life meaning. If it's true, then why do people struggle to find meaning in their lives? Why don't they just go back home and die? Anyway, if you look back at history, humanity has always been about going beyond what our bodies can do. The thing that sets us apart from animals is that we use things like clothes to stay warm when it is cold outside and telescopes to see really far away, which is something that our eyes cannot do on their own. Humanity is about doing things that other animals cannot do and that is what makes us special. Our "essence" isn't our DNA; we desire to overcome. Stealing fire didn't make Prometheus less of a Titan; it made him the ultimate champion of progress.

Why we are The Last Prometheus

To really become Prometheus, we have to figure out the oldest mystery of all: the idea that there is a ghost in the machine. For a long time, people thought that what makes a living body different from a dead body is some kind of special spirit or soul. People believed that life is something that the Gods put into a person by their breath or whisper when they were made. We thought that life was like a gift that made our bodies work, and when we died, this gift was taken away. The idea of Prometheus is connected to understanding what makes us alive, and it starts with understanding that there is no ghost in the machine, just the Last Prometheus trying to make sense of things.

When I look at a cell, I do not see a spirit; rather, I see the cell and how the cell manages energy. The cell is like a thing that uses energy, and the cell manages this energy. The old people used to talk about the "soul". Now we know it is really about Thermodynamics. Life is a small fight against the rules of the universe. The universe wants everything to be messy and cold. It wants everything to be the same with no order. Life is different. Life uses energy to stay tidy and warm. The universe and life are like two opposites. The universe wants entropy, which means things get messy and cold, but life does not want that; rather, it wants to stay organized and warm. It uses energy to do that. Life is like a rebellion against the universe and its rules. When you eat, you are not just putting food in your stomach; you are giving energy to your body. Inside every one of your cells, there are thousands of things called mitochondria. These mitochondria are very busy. They take the food you eat. They make it into a special kind of energy called ATP. This ATP is like a spark that makes your body work. It is what makes your heart beat. Your brain thinks. Your mitochondria are like power plants that use the food you eat to make ATP. This ATP is what gives you the energy to do things. It is what makes you alive. Your body needs ATP to work, so you must eat the right foods to give your mitochondria the energy they need to make ATP. If we can control the energy, we can control the life. We are already learning how to optimize this process - how to make cells live longer, resist stress, and repair damage that was once thought to be "God's will." By seeing life as a chemical machine rather than a spiritual mystery, we gain the power to fix it when it breaks.

We have traveled from the riverbanks of Greek myth to the labs of modern biotechnology. We have seen that the "clay" is actually a complex code and the "spark" is actually chemical energy. The reason this book is titled "The Last Prometheus" is that the cycle of "stealing from the Gods" is finally coming to an end. In the past, humans stole small fires:

- We stole the secret of Healing (Medicine).
- We stole the secret of Communication (Language).
- We stole the secret of Flight (Aerospace).

But now, we are stealing the final secret: Eternity.

We are getting really good at biotechnology, which means we do not have to hope that someone, like a magic person, can give us a few more years to live. We are taking control. Changing when we die. We are getting ready to leave this planet and colonize other planets. We will not be the weak people that nature made. We will be the New Olympians that we want to be. The New Olympians are strong and able to do things that people could not do before. Some people say that if we were supposed to live forever, then nature would have made us that way. But the truth is that nature didn't mean for us to do anything. Nature is a blind process. If we listened to "what was meant to be," we would still be dying of the common cold at age twenty. We have always been a species that defies its "meant to be" status. The most natural thing a human can do is to use their intelligence to improve their condition. Stealing eternity isn't a crime against nature - it is the ultimate fulfillment of human nature.

The cave is now empty. The fire that the first human watched has been replaced by the glow of computer screens and the blue light of sterile laboratories. We no longer need to invent stories about Titans molding us from mud because we can see the process ourselves through a lens. The "Potter" hasn't died because we proved He doesn't exist; the Potter has died because we have taken over the studio. We have realized that the clay is ours, the code is ours, and the future is ours. As we close this chapter on our ancient origins, we move into a world where the "miraculous" becomes "mechanical." In the next chapter, we will see how science is systematically reclaiming the territory of the divine, moving step-by-step through the God of the Gaps. The horizon of what we consider 'impossible' is moving. Every time a biotechnologist sequences a genome or edits a gene, the territory of the supernatural shrinks. We are no longer living in a world governed by the whims of the Gods. We are living in a world governed by the precision of our own tools. It is time to look at the 'Gaps' where we once hid our deities and see what happens when we fill them with the light of science.

2

From Altars to Laboratories

"**God is an ever-receding pocket of scientific ignorance."** — *Neil deGrasse Tyson*

For most of history, the world was a really scary and magical place. When it did not rain, people thought God was upset. If a child got sick, they believed a bad spirit had come into the house. When the sun went away during an eclipse, people thought the heavens were trying to tell them something.

We lived in a world where we always asked why. Why did the storm ruin my crops? Why did my neighbor and I not? We did not know how these things happened. So we tried to figure out who was behind it all. We thought maybe God was behind everything we did not understand. This is what some people call the God of the Gaps. To them, God is the answer to all the things that science cannot explain. God is what fills in the spaces on our maps, the things we do not know about yet. We think of God as being in all the corners where we have not found the answers. When I look at history, I see something that really stands out to me. There is a trend that I think is very important. The dark corners are being lit, one by one. Every time the light turns on, the God who lived there has to move further back into the shadows. I think about this a lot when I consider the field of biotechnology and how it is changing our understanding of the world. The God that I am talking about is not a figure but an idea that is being pushed back by the light of discovery. Biotechnology is helping us to see things that were previously hidden. This is changing the way we think about the world and our place in it. The dark corners that are being lit are the areas of human knowledge, and biotechnology is the light that is shining on them.

Imagine you have an old map of the world from a thousand years ago. In the middle of this map, you can see the village where people live, the roads they use and the hills they know. When you look at the edges of the map, things start to get fuzzy. The people who made these maps did not just leave the edges empty. Instead, they drew things like monsters, dragons and sea serpents. On these maps, they would write: Hic sunt dracones, which means "Here be dragons". The mapmakers put dragons and other scary creatures on the edges of the map to show that they did not know what was there. They were warning people that these areas were unknown and might be dangerous, like the dragons they drew. The maps were like a picture of the world with the village and roads in the middle and the unknown, scary world of dragons, on the outside. The dragons were like stand-ins for things people did not understand. People used the dragons to say that they had no idea what was there, so they just thought it was probably something really strong and magical. The dragons were a way for people to deal with the fact that they did not know everything about the world. The dragons were a mystery.

People liked to think that they were special and powerful, even if they were not really real. The dragons were a way to explain things that people could not explain, and that is why they were so important to people.

In the dim light of our ancestors' fires, the God of the Gaps was the ultimate architect of the unknown. When the stars traced their silent paths across the void, we didn't see the invisible hand of gravity; we saw the fingers of the Divine pushing spheres of light. When the sky fractured with blinding light and the earth trembled under a roar of thunder, we didn't calculate atmospheric pressure; we fell to our knees, trembling before a God who forged lightning in his palms. As the torch of human inquiry pushed further into the dark, the shadows where the monsters lived began to dissolve. We sailed to the edge of the world and found not krakens or cascading abysses, but more water, more land, and the same indifferent laws of physics. We looked into the heart of the atom and the depths of the cosmos, only to find equations where we expected miracles. The God of the Gaps is a deity of diminishing territory. Every time a "how" is answered, the "Who" is forced to vacate another room in the temple of reality. It leaves us with a haunting question for the modern age:

"What happens to a creator when the creation no longer requires a ghost in the machine?"

The God of the Gaps is a god who is losing power. He is like a landlord of a house, but he is slowly losing control of every single room in the God of the Gaps house, getting weaker and weaker.

The History of the Horizon

People have been changing what they think is important for a long time. This has been going on for centuries. Every time someone makes a discovery in science, it is like people are taking back a part of the world that they used to think was controlled by the supernatural. The supernatural is something that people used to think was very powerful. Science has helped us understand that the world is a natural place. Every major scientific discovery is like a step forward for humanity, and it helps us understand the world in a better way. Science is taking the place of the supernatural. That is a good thing for humanity and for science.

From God's Will To Gravity:

Take Astronomy. A long time ago, the sky belonged to the gods. People thought the stars were just holes in the floor of heaven, or maybe they were spirits that watched over us. Then some smart people came along, like Galileo and Newton. They figured out that the moon was not a pearl from heaven; it was actually a rock with apparently big holes in it. They also showed that the planets moved around because of the rules of math, not because the gods decided to move them. So the big mystery of the heavens was solved. We do not need a God to move the stars anymore because we have gravity now. The stars are moving because of gravity. Gravity is what makes the stars move.

From Wrath of God to The Germs:

Then Biology and Germ Theory came along. For a time, people thought the Black Death and other plagues were Acts of God. They believed these things happened because people had done something terribly wrong. People did not ask for medicine to make them better; they asked for mercy. Then we started using the first microscopes. We saw things that we could not see before, Bacteria and viruses, these tiny invisible organisms. When we found out what germs are, the mystery of the plague was gone. We saw that the thing that was making people sick was not a spirit, it was a living thing that we could kill with soap, heat and later with antibiotics. We did not need to think about spirits. The germ was just looking for someone to infect; it did not care what people did was right or wrong. The germ didn't care about your sins; it only cared about finding a host.

From the Seat of the Soul to a Pressure Pump:

For centuries, the heart was considered the most "sacred" organ. Ancient Egyptians believed it was the only organ worth keeping for the afterlife because it held a person's essence. Even in the time of the early Greeks, it was believed that the heart was where the "vital heat" of the soul lived. It was a gap that no one dared to touch.

Then came William Harvey in 1628. By studying the valves and the pulse, he realized something that shocked the religious world: the heart was a pump. It was a muscle governed by the laws of hydraulics. It wasn't a mystical source of "spirit"; it was a mechanical device designed to move fluid through pipes. When we figured out that the heart is basically a pump, the idea of the soul had to change. Today, we have taken this idea far. We can actually take a heart from one person. Put it into another person. We can even replace a heart with a piece of plastic and a battery, which is pretty amazing.

If the heart was thought to be the seat of the soul, what about a person who has a mechanical heart? Does that person have a mechanical soul? The simple truth is that the heart was never really a mystery. It was something that our bodies do, and we did not know how it worked. The heart is like a complicated machine that is inside of us, and it was just a matter of time before we figured out how it works. The heart is a part of our body, and our bodies are like machines that are made up of different parts, and the heart is one of them.

The Harvest: From Sacrifice to Chemistry:

For most of human history, whether or not you ate depended on the mercy of the Gods. Famines were seen as divine punishment. People offered sacrifices, danced, and prayed to "Ceres" or "Osiris" to make the soil fertile. The "Gap" was the mystery of why plants grew.

In the early 1900s, two scientists named Fritz Haber and Carl Bosch closed that gap. They figured out how to pull nitrogen out of the thin air and turn it into fertilizer. This single scientific discovery - the Haber-Bosch process - is responsible for feeding nearly half of the people on Earth today.

The good harvest that people thought was a miracle was not really about making a god happy. It was about the chemicals in the soil, like nitrogen and phosphorus, and potassium. We did not have to say prayers for rain or for the soil to be good for plants. What we had to do was learn what a seed needed to grow. We stopped looking up at the sky for help with hunger. Started looking at what scientists could do in a laboratory to help people have enough food. The good harvest was not a gift from Heaven anymore; it was something that people could make happen in a laboratory.

From the Wrath of Zeus to Static Electricity:

Lightning is perhaps the most famous "Act of God." From Zeus to Thor, every culture saw the bolt of fire from the sky as a weapon of divine judgment. It was the ultimate gap - a power so fast and so deadly that surely only a God could wield it.

Then came Benjamin Franklin and his kite in 1752. By proving that lightning was simply electricity - the same force that makes a spark when you rub wool - he did something incredible. He invented the lightning rod.

Think about what it means that we came up with something like that. For a time, a person could say no to God. We put a metal rod in a church. That was it. We could catch the lightning that people thought was God being angry. It would just go into the ground. It was not scary anymore. We made something that used to be thought of as a thing from God into something that we could understand and predict. We did not stop the lightning from happening; we knew what it would do. The lightning was not special or magical anymore. We took away the power that people thought they had over our lives. The invention of the metal rod on a church steeple was a deal. It was like we were saying that lightning is just lightning, it is not the wrath of God.

The Pattern of Progress:

What do these examples have in common? In every case, the "sacred" became "structural."

- The Heart went from a soul-carrier to a pump.
- The Harvest went from a blessing to chemistry.
- Lightning went from a judgment to a circuit.

I don't start my day by trying to "disprove" God. Science doesn't work that way. Instead, science simply makes the supernatural explanation irrelevant to the process. You don't need to know if there is a creator to build a bridge that stands. You don't need to believe in a soul to edit a gene that causes cancer. The "Gaps" are closing because we are finding that the universe is self-consistent. It follows its own rules. As a biotechnologist, I am standing at the most modern gap: the mystery of life itself. For a long time, people believed there was a "Vital Spark" - a magical energy that separated a living person from a pile of chemicals. They believed this spark was the ultimate gap that science would never bridge. But we have bridged it. We have sequenced the genome. We have found that life is not a mystery; it is information. And just as we reclaimed the stars and the germs, we are now reclaiming the very code of our existence. The goalposts are shifting again. We are moving from asking "Who made us?" to "How are we made?" And once you know the how, you are no longer a subject of the mystery. You are a technician of reality.

The Biological Gap: The Genome (The Final Fortress)

For a long time, we have been learning about the stars and the Earth. We have been figuring out how things work. There is one thing that we still do not understand - Life. The scientists who are not afraid to try new things and do not believe in old stories about the sun and lightning they still think that living things are special. They have a theory called Vitalism. The scientists who believe in Vitalism think that life is different from things. Life is like a mystery that we still need to solve. The mystery of life is what makes living things, like animals and people, different from nonliving things, like rocks and water.

People used to think that living things had an energy inside them called the Vital Spark. This energy was like magic. It did not follow the rules of physics and chemistry. Some people said that you can build a machine like a steam engine using iron. You can never build a heart using simple parts because it is missing something important - the Breath of God. This was a problem that people could not explain. As we started to take a look at the fortress began to fall apart, we saw that the fortress was not as strong as it seemed at first, and the fortress started to break down.

The Death of the Vital Spark:

The big change happened because of something that was discovered. This discovery changed the way we think about our lives. It was the year 1953 when James Watson, Francis Crick and especially the X-ray data from Rosalind Franklin showed us what the DNA double helix looks like. The DNA double helix is really important because it is the DNA that makes us who we are. James Watson, Francis Crick and Rosalind Franklin did a job figuring out the structure of the DNA double helix. This wasn't just a discovery in biology; it was a death blow to the idea of the supernatural "Spark."

What we found was not a spirit. It was not a ghost. It was actually a chemical ladder. We figured out that the Breath of God was really a series of bases, and these bases are Adenine, Cytosine, Guanine and Thymine arranged in a specific order. For the time we saw that life does not need a miracle to work properly. Life needs Information to run the machine, and this Information is what makes life work. The chemical ladder and the nitrogenous bases, like Adenine, Cytosine, Guanine and Thymine, are what give life the Information it needs.

I think this is the moment when the secret of life, which people call the "Potter's Secret," was figured out. We found out that a person is not just something made from clay and spirit. A person is actually like a program that is based on biology. We are, like a set of instructions that have been repeated and passed on for billions of years. When we finally understood that life is based on a code, the special thing that makes us alive, which people call the " Spark," was no longer a mystery. Instead, we started to see how our bodies work in an efficient way, like a machine, and this is all because of molecular biology and the "Potter's Secret" that was discovered. The "Vital Spark" is now we know that our bodies are just following the instructions of molecular biology.

From Sacred Mystery to Digital Data:

Something big changed in that moment. Before people figured out the genome, when a child was born with a problem, it was really sad. People thought it was a test from God. What happened inside a woman's body when she was pregnant was a mystery that only God knew about. The genome is, like a secret code, and before that code was cracked, the genome was not understood.

When we entered the age of the genome, the thing that was a mystery to us became like a Digital Library of the genome. The genome was now something we could look at, and the genome was like a book that we could read. This book had all the information that we could use to understand the genome.

We used to think of our bodies as special and hard to understand. Now we think of them like computers. If a computer has a problem, we do not just hope it gets better. We look for the part, and we fix it. We think about the body in the same way. When someone has a disease like Cystic Fibrosis or Sickle Cell Anemia, we do not think it is because something a higher power wanted. We think it is because of a mistake in the body's code, a coding error, in the human body. We try to fix the coding error in the human body. We see a letter that was written incorrectly by the hands of evolution. The evolution made a mistake when it wrote this letter. This letter is an example of the mistakes that evolution can make with its blind hands.

The transition from "Breath" to "Code" is the most important step in the journey of the Last Prometheus. It is the moment we realize that the "Word" used to wake the Golem isn't a holy secret - it is a chemical sequence. And if it is a sequence, it can be read. And if it can be read, it can be edited.

The End of Biological Innocence:

When we figured out the Genome, something changed. We lost the thing that made us innocent in a way. Now we can not say that we do not know why we are the way we are. The Genome is like a book that tells us about ourselves. We can not say that only God knows why we are like this because now we have the Genome that gives us these answers. We have the Genome, and it tells us a lot about ourselves, so we can not hide behind the idea that only God knows, because the Genome is like a map that shows us who we are.

We have figured out the three billion letters that make up the blueprint. We now know which genes are in charge of the color of our eyes, the strength of our bones and how our brains deal with fear. We have taken back the human Source Code from the area that was thought to be controlled by a power over the human Source Code.

This change is not easy to deal with. When the Gap was still open, it was simple to point fingers at the Gods when things were not perfect. We would say things like "The Gods made me this way" or "Dying's just what happens because that is how nature works". But as the Gap closes, the responsibility shifts. If there is no "Potter" making the choices, then the choices belong to us. I often realize that we are the first generation of humans who are no longer "accidents" of nature. We are the first generation that has the power to look at the "Sacred Blueprint" and say, "This is not good enough." We have moved from being the product of the lab to being the technician in charge of it.

The Final Frontier: The Living Machine

Today, the Biological Gap is almost filled. We are now making life from scratch in dishes. We are also making tissue with 3D printers. The Biological Gap is getting smaller because we are using things like CRISPR to get rid of the things and add good things to it. We are using CRISPR to delete code and insert good code into the Biological Gap.

The God of the Gaps is gone from our bodies. He does not live in our heartbeat anymore; he does not live in our breath. He does not live in our genome. The map of our body is almost finished. Every dragon that we were scared of in our biology has been replaced with something like a protein or an enzyme, or a nucleotide. The God of the Gaps used to be in all these things. Now we know what really makes our bodies work. The genome, the heartbeat and the breath are all things that the God of the Gaps used to be a part of, but not anymore.

The thing that is still on our minds is what we will do with this palace now. We have the code. We know that there is no Vital Spark that will come and save us or stop us. So how far are we willing to go? If we are "information," can we make this information live forever? If we are a program, can we upload this program all the way to the stars?

The goalposts have finally changed for the time. Now we are not trying to figure out the mysteries of the universe; we are getting ready to change them, to rewrite them.

We are finally done figuring out the genome. Now we are standing in a space. This space is what we call the Moral Gap. The Moral Gap is an issue that we need to think about. We have to consider what the Moral Gap means for us and how we can fill this space. The Moral Gap is a problem that affects the genome and our understanding of it.

For a long time, our sense of "Right and Wrong" was tied to the idea that our bodies were sacred property owned by a Creator. We were told that "natural" was good and "artificial" was bad. But if the "natural" world is just a series of chemical accidents and viruses, does that moral system still hold up?

If there is no "Potter" to tell us what is right, then we must decide for ourselves. In the next section, we will explore how the closing of the biological gaps is forcing us to build a new kind of ethics - one that isn't based on ancient myths, but on the responsibility of our own power.

The Moral Gap: The End of "Natural Law"

We wind up standing in a fresh, chilly, vacant gap as the biological barriers are closed. This is the Moral Gap.

For millennia, we inherited our idea of Right and Wrong rather than choosing it for ourselves. We were informed that a Natural Law - a heavenly order to the cosmos - that we were not permitted to alter existed. Natural is good, and artificial is bad according to this moral code. We were informed that it was somehow sacred to age, suffer, and die as God (or Nature) had meant for us to do. However, I view the Natural Law as a falsehood.

Our romanticized view of nature helps to cause the Moral Gap. We like to see nature as a mother: wise, kind, and caring. But in the lab, nature is seen as it really is: a blind, merciless, and uncaring meat-grinding device. Is it Natural Law if a kid is born with a genetic anomaly causing their skin to blister upon contact? According to Natural Law, should a virus kill millions of people by spreading from an animal to humans? If Natural is Good, then we should stop using heart monitors, we should stop using insulin, and we should let every infection run its course.

I came to see that we constructed the Natural Law as merely a barrier for ourselves since we lacked the knowledge to break free. We referred to it as sacred since we had no other option other than to suffer through it. But the cage door opened the instant we invented the vaccine, the antibiotic, and the gene editor. Understanding that Nature is not a moral compass but rather a technical beginning point helps one to see the Moral Gap.

We had someone to blame when the God of the Gaps took over our health area. We would say, God has a plan, or it was their time if someone died young. This served as a mental barrier. It shielded us from the horrible truth that life is fragile and quite frequently unjust. But when we recovered the genome, that barrier broke apart.

If we have the capacity to remove a sickness and we choose not to, who is accountable for the misery that results? We are no longer able to shrug our shoulders and point toward the sky. The Moral Gap is human agency's weighty burden. As the Last Prometheus, we are learning that every defect in the human machine is a problem that we have the power - and so the responsibility - to solve if the design is not guided by a divine hand.

Natural Law ending signals the start of Technical Ethics. We no longer wonder, Is this what God meant? We ask instead:

- *Does this lessen suffering?*
- *Does this increase human possibilities?*
- *Does this increase our resilience?*

You are playing God is the most often used phrase for maintaining the Moral Gap open. A biotechnologist with agnosticism will find no sense in this sentence. We have been acting like God since the first caveman grabbed a firebrand - if playing God entails using intelligence to go beyond the restrictions of our birth. Looking back at the history of the horizon, we find that every significant advance in human knowledge was formerly attributed to playing God.

- *Anesthesia for a woman during labor was formerly considered a transgression of God's order of suffering.*
- *Organ transplants were formerly regarded as compromising the body's sanctity.*
- *People referred to IVF (In-Vitro Fertilization) as playing God with the mystery of conception.*

These are currently typical medical operations. The holy boundary is a ghost that vanishes once the technology is safe and reliable. The phrase playing God has lost its sting because we have shifted the goalposts so often. Those who fear the rate at which the gaps are shrinking just use this term.

The goalposts have been taken from the field. We have brought back the stars, the microbes, the crops, and the code. Natural Law has been debunked, and we have discovered that Potter was never there to keep us from grabbing the equipment. But this is not how the tale ends. This is the conclusion of our innocence.

This chapter has been devoted to examining how science asserts control. However, the following chapter will examine how we actually apply it. Our forefathers will be seen not as victims of the God of the Gaps but rather as Unwitting

Alchemists. We will find that what we believed we were praying was really a basic sort of biotechnology. We were meddlers before we became scientists. We formerly had the farm before the lab came along. And before we became the Last Prometheus, we were the students of the Earth, gradually learning how to change the world around us long before we even knew what a gene even was.

Let's take a look at the statements that have evolved under human supervision:

- **Then**: ***"Some things are meant to be a mystery."***
- **Now**: "Mystery" is just a word for "not yet researched." Every "mystery" we solved has led to a better life for our species.
- **Then**: ***"We are moving too fast; we shouldn't cross these boundaries."***
- **Now**: Boundaries are not divine; they are historical. The boundary of today is the baseline of tomorrow.
- **Then**: ***"Nature's design is balanced; if we change it, we will destroy ourselves."***
- **Now**: Nature is not "balanced"; it is a constant war of survival. We aren't destroying a balance; we are finally tipping the scales in our favor.

3

The Unwitting Alchemists

"**Agriculture is the most fundamental of the biological technologies."** - *Freeman Dyson*

For hundreds of thousands of years, people saw themselves as subservient to nature. We were the Hunters, a name that seemed strong but was really a desperate cry for help to show how limited we were. To be a hunter is to be a scavenger of luck. You go along with the crowd; you do not lead it. You discover the berry bush; you do not design it. We were visitors at a table we did not prepare, living off the leftovers of an uncaring forest. Then, everything changed. We went above the cave. This was a slow, quiet change from being the subjects of nature to being its managers rather than a sudden breakthrough.

We started to see that the world was a system of repeatable patterns rather than a fixed miracle. We came to see that if we planted a seed in the loose dirt close to our camp, a plant sprouted. Should we remove the weeds, it would improve. This marked the beginning of the Accidental Laboratory. The soil, the rain, and a recent, compulsive care to detail comprised our laboratory equipment. We started viewing the forest as a workplace rather than as a sanctuary of gods.

A historian could mention the wheel, the steam engine, or the stone axe as the most significant innovations in human history, but I contend that the most groundbreaking innovation ever created was not made of stone or iron. The Selective Seed it was. While the selective seed let us alter the very nature of the tree, the stone axe let us change its form. Think of Teosinte, a grass that corn came from in the wild. Originally, it was useless to people - a little, hard-shelled cob with just a few kernels that might shatter your teeth. The plant was a work of art in survival, yet it failed for the starving person. Then, roughly 10,000 years ago in Central America, the first Unwitting Alchemists launched their first big Bio-hack.

They lacked CRISPR. They lacked sequencers. They possessed something equally as strong: Selection. They observed that occasionally a Teosinte stalk would yield somewhat softer kernels or a somewhat bigger cob. In the Natural Law of the wild, these changes were weaknesses; They simplified animal consumption of the plant. To humans, however, they were a goldmine. Our ancestors were doing Directed Evolution when they saved certain seeds and planted them again. Reaching into the plant's genetic stream, they were steering its flow toward human usefulness. They changed a worthless weed into the huge, golden ears of corn that feed the world nowadays over thousands of years. This was the first time a species adopted another living being's Source Code and adapted it to fit their own needs; it wasn't farming as we know it.

People decided to make a harvest for the first time instead of waiting for the Gods to bring one. Scarcity helps to characterize the wild. Left to its own devices, nature is not ready to support eight billion people. It aims to keep life and death in a terrible, back-and-forth equilibrium. If you depend on Nature, you depend on the uncertainty of the hunt and the chance of rain. Our forebears declined to take that chance. They came to see that Survival was a project rather than a gift. Domesticating plants was, in effect, telling the cosmos: Your design is unsatisfactory. Your "Natural Law" will cause you to starve. We're going to write an improved law. This was the most audacious display of Promethean rebellion. To keep fed, we didn't only steal fire; we stole the actual reproductive cycles of the earth.

This understanding fundamentally altered human nature. We went from thinking Supplication, which is praying for the forest to be nice, to thinking Sovereignty, which is making sure the field is useful. The wonders of development and life were not strange favors from a god but rather biological mechanisms that could be improved, scaled, and regulated. Through the Accidental Laboratory, we learned that the Earth is flexible. It made us understand that Life is a raw material - like clay - which the human will can mold. We were the builders starting to create a world above the ground, not only animals living in the wild.

Why is this the foundation of Biotechnology?

- **Observation over Superstition**: We looked at the seed, not the sky.
- **Selection over Chance:** We chose the traits we wanted, rather than accepting what we were given.
- **Utility over "Natural Purpose":** We decided that a plant's

purpose was to feed us, not just to survive in the wild.

The Wolf at the Hearth: The First Genetic Amendment

If the domestication of the wolf represents our first victory over the will, then selective breeding was our first triumph over the soil. This was the turning point when people started to control their behavior and biology on a dangerous level instead of just counting calories.

For the old Hunter, the wolf was a terror in the dark, a rival for food and a danger to life. Natural Law controlled this animal, under pack and moon guidance. But then, in a series of events that would become the most successful biological experiment in history, the Unwitting Alchemists did a feat of trans-species engineering that would later produce everything from the Great Dane to the Chihuahua.

This was biological subversion, not just the work of taming one animal. We didn't just catch a wolf; we charmed a species. It started with the leftovers. Less violent wolves - those with somewhat lower flight-or-fight thresholds - scoured near human fires. These kind wolves stood out in the brutal reality of nature. But they discovered a fresh niche close to people. We started to look for wolves who could understand human gestures, wolves who didn't freak out when people got close, and wolves who thought of the human as the new Alpha.

I view this as the first occasion when we hacked a neurological system. Choosing for lower adrenaline levels and greater sociability meant we were changing the physical makeup of the dog's brain. We were breeding out the wild and breeding in the loyal. We were transforming a competitor into a Biological

Tool.

Though they knew Phenotypes - physical appearances, they had no idea what a nucleotide was. They realized that crossing two quick dogs would probably produce quick offspring. If you bred two dogs that were good at pointing at prey, the offspring inherited that instinct. This is Phenotypic Selection, and it is the bedrock of all biotechnology. Through this process, we performed a radical "Edit" on the wolf's anatomy:

- **The Skull:** We shortened the muzzles and shrunk the teeth.
- **The Coat:** We selected colors that nature would have used to mark a target, but we used them to mark a friend.
- **The Instinct:** We took the hunting drive of the wolf and "paused" it. A sheepdog has the drive to stalk and chase, but the "kill" command has been edited out of its behavioral code through thousands of years of human selection.

Many people think of nature as being superior in romantic terms. The dog, though, is proof that artificial often works better. The wolf is a dying species in the wild, fighting for room in a world getting smaller. However, the dog we created, Many people think of nature as being superior in romantic terms. The dog, though, is proof that artificial often works better. The wolf is a dying species in the wild, fighting for room in a world getting smaller. However, the dog we created, Canis lupus familiaris, is among the most successful mammals on Earth. They attained some sort of biological immortality by connecting their DNA to ours.

The awareness here is quite deep: Life is not some immutable, holy thing. It's a flexible, changeable substance.

There is no limit to what we can change if we can transform the top predator of the forest into a creature that sleeps at our feet and understands our spoken words. The proof of concept was the Wolf at the Hearth. It demonstrated to our forefathers that the beasts of nature need not frighten us. We might bring the monsters into our homes, rework their goal, and adopt them. We were now the editors, writing the story of life to include us as the main characters, not only observers of the Sacred Blueprint but also working under the faint light of the campfire.

Why this is a Biotechnology fact:

- **Neurological Engineering:** We selected for neurochemical changes (lower cortisol) long before we knew what those chemicals were.
- **Morphological Plasticity**: We exploited the dog's highly "mutable" genome to create hundreds of specialized "tools" (breeds).
- **Symbiotic Hacking:** We created a biological dependency between two species that was entirely human-made.

The Resurrection of the Grain: Engineering the Engine of Civilization

The Grain promised us a future; the Wolf offered us safety. Grasses are made for survival, not for storage in the wild. They aim to break apart - to scatter their seeds at the slightest breeze so the following generation may grow from the ground. This is the ideal design for a natural world, but a tragedy for someone trying to create a city.

Arguably, the biggest biological recovery ever recorded is the change from wild grasses to cultivated wheat. It shows the time we converted the Breath of Life into a predictable commodity. Natural process for wild wheat (Einkorn) was a fragile rachis. The stalk would break once the seed was ripe, spreading the grain across the ground. This meant a hunter-gatherer had to race the wind to locate food. Unwitting Alchemists then observed a mutation. Once in a while, a stalk showed a genetic fault that let the seeds adhere to the plant even after they were mature: a non-brittle rachis. In the wild, this mutation was a death sentence; the seeds couldn't spread, and the line would end. But to the human observer, this was a Bio-hack of monumental proportions.

Choosing just the damaged plants - the ones that kept their seeds - we basically destroyed the plant's capacity to live without us. Yuval Noah Harari most famously observes in his book Sapiens: A Short History of Humankind:

> *"We did not domesticate wheat. Wheat domesticated us."*

Harari contends that we became slaves to the plant's needs - clearing fields, carrying water, and pest-protecting - while we assumed we were the masters. From a biotechnologist's viewpoint, though, it was a Co-evolutionary Contract. The grain gave us the extra energy needed to build the first labs in return for a worldwide empire.

Many times, old harvest celebrations are regarded as religious Supplications, in which one prays to a goddess such as Demeter or Ceres for a bountiful harvest. But when you zoom in on the grain itself, you see that the miracle was actually

Applied Genetics. Today's wheat is a hexaploid, having six sets of chromosomes - a biological complexity practically never found naturally. Ancient farmers unintentionally (and then deliberately) crossed several kinds of wild grasses to produce a strong, high-protein super-food. The Altar of the Harvest was built on a foundation of human intervention. We didn't just find food; we manufactured it. Author and biologist Jared Diamond claims in his seminal work "The Third Chimpanzee" that, in terms of health, agriculture was the worst mistake in the history of the human race, but he concedes it was the ultimate technological pivot. Diamond points out that we transformed the landscape into a giant "Biological Machine" designed solely to pump out human calories.

The grain was no longer a wild entity when we arrived in the Bronze Age. It was a Living Tool. We changed its height so it wouldn't fall over in the wind; we changed its husk so it would be easier to thresh; and we changed its nutritional profile to help our species' growing brains. From a biotechnologist's perspective, the wheat field is the original Large-Scale Bio-Reactor.

We were no longer hunters searching through what nature permitted. We oversaw genetic output as managers. We had come to see that staying alive required no permission from the Gods if we knew the seed's laws. We had discovered the initial Natural Law loophole: You determine the fate if you regulate the reproduction.

The Resurrection of the Grain was evidence that people could change the surface of the earth to suit their needs. From the fire of our intellect and the clay of the Earth, we had produced a biological engine meant to drive the voyage to the stars.

Why this is a Biotechnology fact:

- **Polyploidy Engineering:** Humans selected for plants with multiple sets of chromosomes for higher yields long before we knew what a chromosome was.
- **Selection for Recessive Traits:** The "non-shattering" trait is genetically recessive; it only survived because humans protected and propagated it.
- **Biological Dependency:** Modern wheat is so "over-engineered" for human use that it can no longer reproduce in the wild without human intervention.

The Magic of the Invisible: Fermentation

Our forefathers were already taming the first microbes long before we knew that life might exist at a level undetectable to the human eye. Though they lived in a society where the border between chemistry and magic was thin, they were the first microbiologists in the world. This was the age of Fermentation, the ancient practice of using biological ghosts to turn the ordinary into the extraordinary. From a biotechnologist's perspective, an old clay amphora of wine is the first bio-reactor in the world rather than a religious artifact. Our ancestors had discovered how to provide a regulated environment where yeast could grow and carry out a sophisticated metabolic activity.

The Unwitting Alchemists were domesticating something far smaller, Saccharomyces cerevisiae, while the hunter-gatherer was busy taming oxen and wolves. They just knew that if you left wet grain or grape juice in a particular jar, it started to bubble and breathe; they didn't realize it was a fungus. Michael

Pollan looks at how plants and bacteria use human desires to guarantee their own evolutionary success in his book "The Botany of Desire." He points out that co-evolution is our interaction with fermenting bacteria. We gave the yeast a vessel and sugar; in exchange, the yeast offered a means of purifying water and conserving calories.

Our forefathers were doing the first Seed Cultures when they made a fresh batch of bread or beer using old dough or the sediment from a past brew. Long before Louis Pasteur established that microbes existed, they were choosing for the most efficient, best-tasting strains of yeast and bacteria, so altering the microbial scene of their homes.

Ancient people thought of fermentation as a wonder. Through a divine process of decomposition and effervescence, you transformed sweet juice into a liquid that could change the mind, heal wounds, and stay fresh for years. Dionysus in Greece, Osiris in Egypt, or Ninkasi in Sumer were regarded as gifts from the gods.

A modern pharmaceutical factory where we create insulin using genetically engineered bacteria is the direct descendent of an old Egyptian brewery. The Magic of the Invisible taught us three critical lessons that define modern Biotechnology:

Sterilization: Ancient brewers knew that if the "wrong" spirit (bacteria) entered the jar, the batch was ruined. They were practicing early aseptic techniques.

Optimization: They learned that temperature and sugar levels changed the "strength" of the spirit. They were the first to understand Metabolic Flux.

Preservation: They realized that biology could be used to fight biology—using the acidity of fermentation to prevent the growth of toxic rot.

The intellectual realization here is that we didn't need a God to bless the wine; we needed the right temperature and the right strain of yeast. We were "The Last Prometheus" even then, reaching into the microscopic world to steal a different kind of fire - the fire of Chemical Transformation.

Why this is a Biotechnology fact:

- **Microbial Domestication:** Saccharomyces cerevisiae has been so heavily selected by humans that modern brewing strains are genetically distinct from their wild ancestors.
- **Biological Preservation:** Fermentation (producing lactic acid or ethanol) was the first "Bio-security" measure, creating environments hostile to human pathogens.
- **Metabolic Engineering:** By controlling the inputs (malt, honey, grapes), humans were essentially "programming" the output of microbial colonies.

The Living Tool: Domestication as Technology

If fermentation lets us capture the unseen, horse and ox domestication let us overcome the material. The Unwitting Alchemists started to see animals as Biological Engines rather than just walking flesh on this day. We were now redesigning the anatomy and bones of the animal kingdom to drive our development, not merely changing a plant's seeds or a microbe's waste. From a science guy's viewpoint, this seems like the first mechanical bio-engineering iteration. We adapted the pure force of nature to fit the human wheel.

The Ox came first, before the steam engine or the electric motor. An ox's strength in the wild is meant to help it protect itself and show off its mating displays. But the instant we attached a yoke around its neck, we turned that biological energy toward a human purpose: the plow.

Jared Diamond argues in his seminal work "Guns, Germs, and Steel" that the most crucial element for the growth of Eurasian civilizations was the domestication of great mammals. He says that among the several thousand big wild mammal species worldwide, only a small number - the Major Five (sheep, goat, cow, pig, horse) - possess the particular biological qualities needed for human utilization. Diamond observes that we were choosing for a rather particular Nervous System Profile, not only domesticating these creatures. We wanted animals who:

- Followed a hierarchy that let a human become the alpha.
- Had a flight response that did not entail immediate panic.
- For most wild species, a big biological obstacle is breeding in captivity.

We basically did a Neuro-Chemical Edit on these species, picking those with less adrenaline and a better tolerance for being locked up.

The modern dairy cow is among the best examples of Unwitting Alchemy. A mammal in the wild makes only enough milk to feed its young. The milk ends once the calf is weaned. This is the Natural Law of biological efficiency. We broke that law, though. We developed a Biological Glitch across thousands of years of selective breeding. We bred cows to keep producing milk far beyond the requirements of a calf, and we picked for cows that did so. A maternal reproductive function was turned into a Continuous Production Line.

As Vaclav Smil points out in "Prime Movers of Globalization", we were "burning" grass through the body of an ox to turn the soil. We had turned the animal into a living machine that converted low-grade biomass into high-grade kinetic energy.

This change marks a significant turn in human morals. We concluded for the first time that a creature's biological purpose - to live and reproduce for itself - came second to its human utility. We are learning to negotiate a set of rules that nature has rather than breaking some inherent plan. We turned the horse and the ox around rather than stealing their speed or strength. Biology turned out to be modular. By breeding a bull's strength module with a human-raised calf's docility module, you may produce a tool that never existed in the wild. Treating life as a tool, the Unwitting Alchemists set the scene for modern biotechnology. A bacterium can produce a medicine for us. Today, we are doing just what our forefathers did with the ox: we are giving a living thing a new, human-defined function. From being passengers on the Nature Express, we turned to conductors, determining precisely where the biological energy of the Earth should be spent.

Why this is a Biotechnology fact:

- **Behavioral Genetics:** Early domestication was the first large-scale selection for specific behavioral alleles (traits) in mammals.
- **Energy Conversion:** Using animals as "prime movers" was the first successful attempt at harvesting biological

energy for non-biological work (construction, farming).

- **Physiological Redesign:** Through breeding, we altered the skeletal structures of horses and oxen to make them more suitable for bearing weight and pulling loads, effectively performing "slow-motion" structural engineering.

The Alchemists of the Silk Road

Easterners were engaging in an even more advanced kind of biological espionage and engineering as the Fertile Crescent's farmers were hacking oxen and grains. During this time, the Silkworm (Bombyx mori) was very popular. The silkworm was our first Industrial Bio-Factory; the dog was our first partner; the ox was our first engine.

The narrative of silk is not only one of fashion; it is also one of how people took a wild, boring insect and changed it so much that it came to rely on us totally for its survival - a living machine designed for only one, luxurious output.

The ancestor of the silkworm (Bombyx mandarina) is a robust animal in the wild. It hides from predators, it flies, and it thrives in the harsh Chinese seasonal cycles. However, the current silkworm is a distinct being altogether. Over thousands of years of human alchemy, we produced an insect without sight, incapable of flight, and without the innate knowledge to avoid predators. Peter Frankopan refers to silk in his book "The Silk Roads: A New History of the World" as the first worldwide currency. More significantly still, Frankopan underlines from a biotechnological angle that the creation of silk was a state secret kept under the death penalty. Why? The rest of the world could not even begin to understand the extent to which the Chinese had perfected Environmental Control and Selective

Breeding. They had discovered they could maximize the silk glands' protein output by feeding the larvae a rigorous mono-diet of white mulberry leaves. We were running a biological production line, not only growing bugs.

The West assumed silk was a plant or mineral for ages. Pliny the Elder, a Roman historian, believed it to be a type of wool obtained from the Seres' woodlands. The enigma of the animal's life cycle was the Gap here. The Chinese had the Source Code, and they were not sharing. Among the biggest episodes of Bio-Piracy in history is the breaking of this dominance. According to myth, two monks sent by Byzantine Emperor Justinian smuggled mulberry seeds and silkworm eggs out of China in the 6th century using hollowed-out walking sticks. This was the Middle Ages' Promethean flame. They were stealing a Genetic Program when they stole the eggs, not only a commodity. They came to see that with the eggs (the software) and the mulberry trees (the equipment), one could reproduce the business anywhere. For the first time, people understood that an organism's DNA holds its biological riches and that it may be moved about.

The ancient world thought of the shimmering thread of silk as a wonder from the sky. It was a chemical outcome for the Unknowing Alchemist. They discovered that boiling the cocoon at the precisely proper time would allow you to dissolve the sericin, the glue, and unravel a single, unbroken strand of fibroin protein about a kilometer long. As Mark Miodownik explains in his book "Stuff Matters: Exploring the Marvelous Materials That Shape Our Man-Made World", silk is an engineering marvel. It is stronger than steel by weight and incredibly elastic. Miodownik notes that the silkworm is essentially a "living 3D printer."

The intellectual understanding here is that we only need to harvest by manipulating the organism; nature abounds with High-Performance Materials that we need not invent. We came to see that the holy moth was really a source of structural proteins. We simply had to grasp the life cycle of the insect and supply the correct chemical precursors - mulberry leaves - we did not require a miracle to make silk.

The Garden of Curiosities: Early Hybridization and Grafting

While the silkworm was a masterclass in animal domestication, the ancient gardens of the Mediterranean and the hanging gardens of Babylon were the scenes of a different kind of alchemy: Surgical Botany. The Unknowing Alchemists found here that physical fusion could change life as well as breeding. This was the start of grafting, the first time people saw two different living things and decided to combine them.

Long before we had molecular scissors like CRISPR, we had the grafting knife. Ancient Greeks and Romans found out that the two would merge if you cut a branch (the scion) from a good fruit tree and fastened it to the rooted trunk (the rootstock) of a stronger, wilder tree. The xylem and phloem, the circulatory systems of the plants, would combine to produce a biological chimera. Pliny the Elder wrote in his seminal Natural History that he was amazed by trees that produced several types of fruit on every branch. Pears, apples, and plums all grew from the same trunk. From a biotechnologist's perspective, this was the first platform technology. Our forefathers had come to understand that a plant's hardware - the roots - could be distinguished from its software, the fruit-bearing branches.

The "Garden of Curiosities" was the first place where humans prioritized Aesthetics and Taste over biological survival. In the wild, a tree that produces massive, sugary peaches is a failure; it attracts too many pests and breaks its own branches under the weight. Nature wants "balanced" fruit; humans want "excessive" fruit. In the book "The Invention of Nature" by Andrea Wulf, it is said that ancient gardeners showed that the Natural Order was just a hypothesis by producing hybrids like the Citrange or particular roses.

Species are not immutable containers. They are changeable. If you may grab a branch from a sweet orange tree in India and graft it onto a sour lemon root in Italy, you will create a life form that God or Nature never intended. From gardener to Biological Architect, you have changed.

Perhaps the most famous Bio-rescue in history happened in the 19th century when the European wine industry was nearly destroyed by a tiny insect called Phylloxera. The natural European vines had no defense. The Alchemist's solution? They grafted the delicate European vines onto rugged American rootstocks that were naturally immune to the pest. This saved the world's wine, but it also proved a cold truth: The "Sacred" can only survive if it is supported by the "Functional." We didn't need a miracle to save the vineyards; we needed a surgical intervention between two different genomes.

Why this is a Biotechnology fact:

- **Chimerism**: Grafting creates a "bigenomic" organism where two different sets of DNA coexist in one body - the earliest form of physical genetic engineering.

- **Asexual Cloning:** This was the first time humans realized they could bypass natural reproduction to create identical copies of a superior organism.
- **Vascular Integration:** Ancient grafting was the first successful attempt at human-mediated "organ transplantation" in the living world.

The Plague Doctor's Wisdom: Early Immunology

While gardeners were fusing trees, others were looking at the human body's ability to "update" itself. This brings us to Variolation- the ancient ancestor of the vaccine. Variolation was a historical, pre-vaccination method of immunizing against smallpox by deliberately exposing healthy individuals to live smallpox material (pus or scabs) from infected patients

Long before Edward Jenner and the smallpox vaccine, doctors in China, India, and the Ottoman Empire were carrying out a horrific act of Unwitting Alchemy. They would grind the dried scabs from a smallpox victim into a powder and blow it up the nose of a healthy person or scratch it into the skin. This was the first occasion people came to see that the body had memory. They knew about the Algorithm of Immunity even if they didn't know about T-cells or antibodies. A little of the enemy now keeps a deadly invasion off for tomorrow.

Steven Johnson investigates in The Ghost Map how people started to fight unseen diseases by means of data and observations. The riskiest Bio-hack was variolation. If you gave too much, the patient died; if you gave exactly enough, you programmed their immune system for life.

My personal understanding here is that Purity is just a fiction.

We have to interact with the dirt of the earth in order to be healthy. We discovered we could train our biology the way we train a dog. Once more we were the Last Prometheus, grabbing the plague's fire to guard ourselves from the flame.

The Sacred Cat: Bio-Engineering for the Afterlife

The ancient Egyptians turned their attention to the ultimate biological barrier: decay, even as the farmers and brewers were busy hacking the living. Death to them was a technical issue needing a chemical remedy rather than an end. The mummification chamber was their laboratory, and they were the first tissue engineers in the world.

Natural Law, the Egyptians discovered, said a body had to return to the earth. The universe's clean-up team is bacteria, moisture, and enzymes. To halt them, the Egyptians had to complete a total Biological System Override. They dried the body with natron, a natural salt that stops the breakdown of cells that starts when the heart stops. Famous Egyptologist E.A. Wallis Budge writes in his book "The Egyptian Book of the Dead" that the voyage of the soul depended on keeping the physical body intact. Underneath the religious ceremony, though, was a Biochemistry lesson. They used resins, cedar oil, and bitumen as antimicrobial barriers, much like we use preservatives in food or medicines today.

The Sacred Cat marks the first time people studied anatomy to conserve rather than to kill. They were the first to realize that several tissues decay at varying rates when they removed the brain through the nose and the organs through a side incision. They were tracing out human biology's shelf-life. A fascinating fact is that Egyptians mummified millions of animals, including cats, crocodiles, and ibis, not only people. Years later, we discovered a mummified bull so well-preserved by Alchemist methods that its muscle fibers were still visible.

We discovered that Eternity is not a present from the gods; it is rather a consequence of good pH balance and moisture management. If you know the chemistry of the cell, the Egyptians showed that one can slow down the biological clock.

From the Farm to the Pharma: The Bridge to Today

We have been Unwitting Alchemists for ten thousand years. We have bred poodles from wolves, transformed natural grass into a worldwide wheat empire, and employed invisible fungus to create the basis of civilization. But the time of innocence is past. Stewart, we have progressed from the Slow Layer of conventional farming, where it took centuries to create a new breed of cow, to the Fast Layer of biotechnology, where we can alter a genome in a weekend.

We are not waiting for a fortunate mutation in the field now. We are producing mutations on purpose. While the ancient Alchemists dealt with the Result of the code - that is, the plant or animal - we labor with the Code itself.

From our forefathers, the main lesson is that human beings are basically biological architects. We were never intended to leave nature untouched. We dedicated ourselves to the Promethean path from the instant the first person picked up a selective seed. You know, if you brought a contemporary organic ear of corn back to a farmer from 5,000 years ago, they would consider it a mystical object. To them, thousands of years of intense genetic tinkering have produced our natural food.

As we move into Chapter 4, we leave the "Unwitting" behind. We are no longer the students of the Earth, guessing at the secrets of the seed and the spirit. We are the Architects. We have moved from the campfire to the clean room, from the clay jar to the CRISPR kit. The mask of the alchemist has been removed, and the goggles of the biotechnologist have been put on. We are finally ready to do intentionally what our ancestors did by accident: We are going to rewrite the future of life.

4

The Code-Breaker

"We used to think our fate was in the stars. Now we know, in large essence, it is in our genes." - *James Watson*

For ages, the idea of heredity was the ultimate black box. Children looked like their parents; we knew sweet fruit would grow from a sweet apple seed. Still, the "How" was spectral. To the ancient world and even to the scientists of the early 19th century, the passing of life was a Sacred Whisper, a magical force that moved through the blood and carried the essence of the ancestors. This was the age of the Black Box, when the inner workings of life were deemed too difficult or possibly too sacred for the human intellect to understand. But the Last Prometheus became sick of analogies when the 20th century approached. We changed our search from a ghost to a molecule.

The change started when we discovered that heredity was a physical thing rather than a fluid essence or a holy spark. Biology before the 1950s mostly involved stamps: collecting, naming, and classifying. Though we could define a butterfly's wings or the petals of a flower, we could not perceive the Architect behind them. But the breakthrough occurred when we understood that a physical structure following the principles of chemistry and physics kept the instructions for life. In his landmark book "What is Life?", the physicist Erwin Schrödinger suggested that the heredity carrier should be an aperiodic crystal, a complex, non-repeating structure capable of storing enormous volumes of information.

This was the point at which the mystery started to die. This was the moment the "Mystery" began to die. If life were a crystal, it could be measured. If it were a molecule, it could be mapped. We moved from the abstract to the concrete. We realized that the "Soul" wasn't what made us unique; it was the specific arrangement of atoms in our cells. I regard this change as the most significant event in human history. When you believe that life is a divine mystery, you are a spectator. When you realize that life is a chemical script, you become a Reader.

Building a person, we learnt, calls for no magic at all. Only information exists. The Script of Life, like a play penned on paper, may be read, dissected, and ultimately criticised. We came to see that a fool and a genius, or a healthy man and a sick one, differed not by fate or karma. It came down to syntax. Author Horace Freeland Judson vividly depicts this metamorphosis in his book "The Eighth Day of Creation." he sees molecular biology's birth not as the discovery of a new species but rather as the discovery of a fresh language. Revealed to be a set of chemical instructions was the mystery of the ages. The God of the Gaps had been hiding in the smallness of the cell, and we had at last constructed the microscopes to locate Him.

The Black Box of life had been opened by the mid-20th century. It was full of a long, coiled molecule holding the history of 4 billion years of evolution; it wasn't empty and it wasn't filled with light. Many researchers thought proteins were the carriers of genetic information before DNA was discovered because they were more complicated. They believed DNA was too simple and boring to contain the secrets of life. As it turns out, nature likes the grace of a simple code above the messiness of a sophisticated one.

The "Death of Mystery" was the prerequisite for everything that followed. We had to stop worshipping the "Spark" before we could learn to control the "Fire." We had moved from a world of "What" to a world of "How," and in doing so, we crossed the threshold into the era of the Code-Breakers.

The Double Helix: The Greatest Heist

The revelation of DNA's structure was a high-stakes, intellectual heist that altered the course of the human species, not only a quiet moment in a lab. While the previous section dealt with the Death of Mystery, this section focuses on the birth of Visual Certainty. In 1953, we at last got to view life's face rather than simply speculating about it.

The race to find the structure of DNA was the "Space Race" of the 20th century. In the running were the famous duo James Watson and Francis Crick at Cambridge, and the legendary chemist Linus Pauling in the United States. But the most important player was the one holding the camera: Rosalind Franklin.

Franklin was a specialist in X-ray crystallography, a method whereby X-rays are fired at a crystallised molecule, and the pattern of their bouncing off is recorded. She took a dry, hazy picture called Photo 51 in 1952. It appeared like a grey X within a circle to the uninitiated. To a physicist, though, that X was a mathematical fingerprint. It yelled one word: Helix. This image was presented to Watson and Crick without her permission or knowledge. That was the smoking gun they required. It gave the precise symmetry and dimensions needed to make a physical model. They did not discover DNA; Franklin's data served as the missing piece that helped them complete the jigsaw. When Watson and Crick finished their model, they went to The Eagle pub in Cambridge and famously announced to a room full of confused locals that they had "found the secret of life." They weren't being hyperbolic; they had found the physical mechanism that allowed 4 billion years of evolution to happen without a single "divine" intervention.

The world came to see that life's secret was a masterwork of Geometry when Watson and Crick at last put together their model, a towering framework of metal plates and cables. Base pairs fit together like a lock and key to keep the two strands together.

Adenine (A) always goes with Thine (T).

Guanine (G) always pairs with Cytosine (C).

The symmetry here helped to clarify the "How" of living. A cell's helix just unzips when it has to split. Every strand then serves as a blueprint for creating a new strand. It was the most sophisticated copier ever found. It demonstrated that life could be passed on by nature without a miracle; it only needed a Complementary Sequence.

For the Last Prometheus, the Double Helix served as the last evidence confirming the universe runs on "Order" rather than "Whimsy". If life were truly supernatural, its core would be chaotic, shimmering, or incomprehensible. Instead, the core of life is a repetitive, mathematical spiral. It is something that can be measured with a ruler and calculated with an equation. This discovery moved the conversation from "Who made us?" to "How are we built?"

Knowing the Double Helix altered our view of ourselves; we understood we were biologically designed rather than holy clay. But with this perspective came a fresh sort of terror. Eventually, we could touch the structure if we could see it. Knowing how the Lock operated, we would most definitely try to create a key.

The heist was complete. We had stolen the blueprint from the safe of nature. Now, the question wasn't whether we understood life—it was whether we were ready to be its master.

The Digital Transition: Life as Data

Biology had a profound transformation as the helix structure was discovered. It started to be about Information Technology rather than tissues and organs. We came to see that our DNA's A, C, G, and T were bits, not only chemicals. They represented the biological operating system's four-letter binary.

Looking at a computer reveals hardware (the screen, the chips) as well as software (the code instructing the chips what to do). People are the same. The hardware is our muscles, eyes, and bones. The Source Code is our DNA. This transformation converted the lab to a data center. We came to see that a virus is basically a bad script that uses our cell phone to make copies of itself. Cancer is a software bug telling a cell to divide excessively.

If life is data, then life is universal. The same four letters that make a person also make a blade of grass, a big white shark, and a colony of bacteria. We are all written in the same language; we are only different apps running on the same planetary server.

Scientists started to question if DNA was the best way to store information in the 1990s. DNA is very stable as opposed to a hard drive that degrades in ten years. Ancient organisms caught in amber or permafrost have had their genetic code effectively extracted and read by scientists. Scientists read DNA from a million-year-old mammoth in 2021. This showed that among all the records in the universe, the Data of Life is the most resilient. We are more than just flesh and blood; we are a Biological Archive that has been carrying on a message continuously since the beginning of time.

Two meters long would result if you were to pull out the DNA from one human cell. If you repeated this for every cell in your body, the string of information you would get would reach from the Sun to the Earth and back more than 60 times. You have a library of knowledge so large that it would take 50 years to complete your own handbook if you attempted to write it out at 60 words per minute, 8 hours daily. We are only starting to locate the Search bar, and we are the most intricate programs ever created.

The Human Genome Project: Mapping the Territory

If DNA were the language, we would require a dictionary. The world started the Human Genome Project (HGP) in 1990. It was like the biological version of the Apollo 11 moon landing. Reading all three billion letters of the human blueprint required a worldwide effort.

The endeavor involved great risk. We lacked the computers and speed required for effective performance at the moment. For thirteen years, it was an Inner Space of our kind adventure. When it was finished in 2003, we had made a map of every gene that makes a human being. This was the last barrier for the Last Prometheus. We had mapped ourselves after having mapped the stars. We went from enigmatic to a Known Quantity. No part of the human body could now conceal a soul or a spirit from the light of knowledge. The Human Genome Project was not just a quiet scientific project; it was a race with cutthroat competition. The government-funded project was challenged by a private business named Celera Genomics, headed by the unconventional scientist Craig Venter. Venter said he could stitch the genome back together with a supercomputer by using a method called Shotgun Sequencing, which is like breaking it into millions of pieces. This would be faster and cheaper. The White House had to step in since the competition grew so heated. Bill Clinton gave a press conference with both leaders in June 2000 to reveal a tie. This competition sped up our technology by decades, demonstrating that people would go to any length to be the first to grasp the Blueprint of Life when the reward is the Blueprint of Life.

The first human genome to be sequenced didn't really belong to one person. It was a composite created from the DNA of numerous unidentified volunteers. Despite the project's $3 billion price tag and thousands of scientists involved, we discovered that two people are 99.9% similar in their genetic makeup. The vast diversity of our world - every race, every height, every talent - lives in that tiny 0.1% of difference. We are a species of practically identical books with only a few thousand misspellings that set us apart.

Breaking the Central Dogma

If the DNA is the Source Code, it is useless if it stays locked inside the safe of the nucleus. Creating a living, breathing organism calls for the reading and carrying out of that code. This mechanism in molecular biology is controlled by a principle so basic that we refer to it as the Central Dogma. The Central Dogma explains the one-way passage of biological data:

DNA → RNA →Protein

Think of DNA as a master architect's blueprint that never leaves the office. To actually build the house, the architect makes a photocopy of a specific page—this is mRNA (Messenger RNA). This photocopy travels out to the construction site, where a massive molecular machine called the Ribosome reads the instructions and begins "printing" a protein. Reading and printing your code proceeds at a speed virtually impossible to picture. At around 20 units per second, your ribosomes can join amino acids into a protein. Millions of these printers are buzzing about within you every second. If your Central Dogma stopped working for even a single minute, your cellular "hardware" would collapse, and life would simply cease to exist. We are held together by a constant, high-speed stream of data translation.

Proteins are the "workers" of the body. They are the hemoglobin that carries your oxygen, the collagen that holds your skin together, and the enzymes that digest your food. The Central Dogma was the moment we realized we could bypass "nature" altogether. If we could write a custom piece of mRNA, we could tell the body to print anything we wanted.

Breaking the Dogma's most well-known actual application came during the 2020 epidemic. Traditionally, vaccines include injecting a section of a virus into the body. Still, mRNA vaccines like Pfizer and Moderna revolutionized things. Scientists sent a Digital Script instead of a virus. They sent a length of mRNA instructing our ribosomes to print the spike protein of the virus. The vaccination factories became our own bodies. Fighting a plague, we first hacked the Central Dogma on a worldwide level and pitted our interior 3D printers against a biological invader, turning our inside.

The Polymerase Chain Reaction (PCR): The Biological Copier

Scarcity is the biggest threat in the Code-Breaker's world. DNA is present in a single drop of blood or a fragment of old bone, but in such tiny amounts that research on it is practically impossible. Imagine trying to read a book, but you only have one copy, it's written in microscopic ink, and it's buried at the bottom of the ocean. We had to rapidly print billions of copies of a particular page in order to examine the code. The Polymerase Chain Reaction (PCR), the biological equivalent of a high-speed Xerox machine, helped to solve this one.

Created in 1983, PCR is a process that uses heat and a specific enzyme to set off a chain reaction. It repeats opening the DNA, copying it, and zipping it back up. The beauty of PCR is its exponential growth.

- In the first cycle, you have 2 copies.
- In the second, 4.
- By the 30th cycle, you have over one billion copies.

PCR was the last Prometheus's best magnifying glass. It let

us turn one shout from our genetic past into a worldwide broadcast. Modern forensics, rapid disease testing, and in-depth ancestry findings all depend on PCR. Without it, none of these would exist. It converted the unseen into the industrial.

Among the strangest tales in science is the discovery of PCR. It didn't occur during company hours in a pristine laboratory. It occurred while driving through Northern California's mountains under moonlight. The chemist who came up with the idea, Kary Mullis, was driving with his girlfriend when the way a chain reaction works suddenly struck him like a lightning bolt. Mullis was a controversial character who famously said that taking LSD helped him see the molecules and picture how DNA replicates itself. Whether or not that's true, he won the Nobel Prize for his Midnight Eureka discovery. He realized that automating the copying process would be possible if he used a heat-resistant enzyme discovered in bacteria dwelling in the boiling geysers of Yellowstone National Park. Using a tool from a monster (extremophile bacteria) to benefit humans was a typical Alchemist action.

PCR lets us converse with the dead. Hikers in the Alps discovered a frozen corpse called Ötzi the Iceman in 1991. It had been dead for more than 5,000 years. His DNA was broken and deteriorated owing to his great age. Scientists can amplify the small bits of leftover code using PCR. They discovered not only his brown eyes and lactose intolerance but also the DNA of the Lyme disease bacteria in his system and even the DNA of his last meal, ibex and red deer. PCR lets us rebuild a whole life from only a few scattered letters left behind on the ice.

Before PCR, a paternity case or a criminal trial usually came down to he-said, she-said or unreliable eyewitness evidence. Molecular Mathematics replaced human intuition in PCR. It revealed that wherever life goes, it leaves a digital trail. In a world where one skin cell can be expanded into a mountain of evidence, there is no such thing as a perfect crime or a secret ancestry. We came to see that the truth is something we can print rather than something we have to believe in.

The Ghost in the Machine: Junk DNA

Scientists expected to discover a large, well-organized library of instructions - a clear plan for every bone, neuron, and enzyme - when the initial Human Genome Project draft was finished. They discovered a wilderness instead. Just around 2% of our three billion A, C, G, and T letters encode for proteins, they found. The other 98% seemed to be nonsense. It seemed absurd, meaningless, and repetitive. Scientists dismissively referred to this large area as Junk DNA for years. It was like we had opened the Book of Life just to discover that most of the pages consisted of arbitrary noise, old commercials, and deleted scenes. But for the Last Prometheus, this desolate land ended up being the most enigmatic and potent component of our nature.

We currently refer to this 98% as the genome's Dark Matter. It's not trash; it's an Evolutionary Attic. Our DNA is a palimpsest, a parchment written over, erased, and stained for millions of years, not a clean record. The scars of every battle our forebears ever engaged in reside within this dark matter. We bear the shattered vestiges of long-ago viruses that infected our forebears millions of years ago and never vanished. We have pseudogenes, which are the ghosts of instructions for traits we no longer need, like the ability to make Vitamin C or the sense of smell of a bloodhound.

You are not really human, which is among the most perplexing results in modern genomics. About 8% to 10% of your DNA is made up of Endogenous Retroviruses (HERVs), which are the genetic leftovers of old viral infections. Millions of years ago, a virus infected an early primate and managed to stitch its own code into the primate's sperm or egg cells. The virus grew to be part of the family tree rather than killing the host. We once believed these were only dead travellers. However, researchers have just found that a particular viral protein called Syncytin is really crucial for the growth of the human placenta. People could not produce live children without this old viral garbage. The exact program enabling us to reproduce is a malicious script from a prehistoric virus. We are basically part-virus.

You have probably come across the fascinating truth that bananas have 50% of their DNA in common with people. Though we clearly don't resemble fruit, this is not a joke; it's serious proof of the Standardized Software of life. 50% overlap happens because the fundamental Operating System functions of a living cell, such as how to convert energy, copy DNA, and build a cell membrane, are the same across all cells. Dark Matter and the basic coding building blocks for cell life stay the same whether you are a piece of fruit or a scientist. We are not a new creation; we are a Premium Version of the same code running the Earth.

Here, the intellectual understanding is that the Conductor is really the Junk.

If the 2% (the genes) are the musical instruments, the 98% (the non-coding DNA) is the Sheet Music. It instructs the body when to switch on a gene, how loudly it should be, and when to stop. This explains why, despite being 99% similar in their coding DNA, a man and a chimpanzee exhibit such variation in look and behavior. It's in the direction, not the equipment. From believing our genome is a basic handbook, we have come to see it as a complicated, multi-layered Symphony of Information. The Ghost in the Machine is a large, data-driven regulatory network we are only starting to understand, not a spirit.

Bio-Informatics: Where Biology Meets Silicon

In the 21st century, the laboratory has changed silently and steadily. The strong odor of bleach and the clanging of glass test tubes have given way to the repetitive drone of cooling fans and the flicker of blue LED lights. Welcome to the age of bio-informatics, the carbon-silicon alliance.

As the Last Prometheus, we understood that knowing the genome (Chapter 4) was only half the battle. One thing is reading three billion letters of A, C, G, and T; another is knowing their meaning. Finding one particular, misspelled word among the 3,000 dense volumes in a library is like searching for one single disease-causing mutation in a human genome. Human eyes are unable to accomplish this. We were seeking a digital brain.

Biology became an Infinite Library when it went digital. Each time someone's genome is sequenced, about 200 gigabytes of raw data are produced. If we were to sequence every person on Earth, the results would go beyond the storage capability of every hard disc now available. This made the coder and the biologist have to merge. An algorithm is now the most effective tool available in a genetics lab; it replaces a centrifuge. We're digging life now, not viewing it. For thousands of years, we have treated the body as a database, looking for invisible connections to us.

The Protein Folding Problem tormented biologists for 50 years. We understood the protein code, but we were unable to forecast its structure. Shape determines destiny in biology. If a protein folds in the wrong direction, it causes Alzheimer's, Parkinson's, or Cystic Fibrosis. Predicting the structure of one protein used to need five years of manual effort from a PhD candidate decades ago. Google-owned artificial intelligence AlphaFold accomplished the unachievable in 2020. Training on the Big Data of known proteins enabled the artificial intelligence to grasp the principles of folding. It predicted the shapes of almost all of the more than 200 million proteins known to science in a single weekend. Overnight, it was the biological equivalent of moving from a hand-drawn map to Google Earth. At last, Silicon knew Carbon's geometry.

Scientists usually don't venture into the jungle when looking for the roots of unknown diseases; they go into the SRA (Sequence Read Archive), a vast public database of genetic snippets from thousands of unrelated studies. Scientists recently searched through this digital repository and found thousands of previously undiscovered viruses practically hiding in plain view. Other labs had mistakenly sequenced these viruses and thrown them out as noise or pollution. Re-scanning these digital rubbish heaps with great computing power reveals fresh life forms never seen before, while we never leave our desks. The first generation of adventurers able to find a fresh species within a Virginia server is us.

Moore's Law is the concept that computers double in speed every two years. We have something much more ambitious in biotechnology: The Carlson Curve. The Carlson curve describes the exponential decline in the cost and the corresponding rise in speed of DNA sequencing and synthesis, serving as a biotechnological equivalent to Moore's Law. DNA sequencing is five times faster than computer chip pricing. It cost $100 million in 2001 to sequence one human genome. You can do it now for less than the price of a high-end smartphone ($600–$1,000). The Language of Life is more easily available than a college textbook since we are seeing the quickest technical collapse in human history.

We are looking back at the first Storage Device, DNA, as our silicon hard drives battle to match the data we generate. Among all the materials in the known universe, DNA is the most dense one for storing information. Inside a shoebox loaded with DNA, you could store the whole internet - every video, every picture, every tweet ever published. The Wizard of Oz and even a brief French movie have already been successfully coded into DNA strands by scientists. We are realizing that the best way to store the future of our digital world might be to write it back into the biological code that created us.

The merger of biology and silicon brings us to a chilling realization: If life is just code, then life can be simulated. We are approaching a point where we don't need to test a drug on a human or an animal; we can test it on a "Digital Twin" - a perfect computer model of your specific genome. This removes the "trial and error" of nature and replaces it with the "precision and simulation" of engineering. We have officially moved from being a part of the "natural world" to being the operators of a Biological Operating System.

The Architect's Threshold

We are currently at the last edge of the Precision Era. Behind us is the history of the Unwitting Alchemist - thousands of years of breeding wolves into dogs and wild grass into wheat, following the slow, disorganized tempo of nature. In this chapter, we have translated the Language of God into the binary code of our supercomputers, read the master blueprint, and broken the locks of the cell.

But a major change is happening right as we stand here. We are not only the Code-Breakers interpreting the past; we are developing the Architects who define the future. We're going over the line between reading and writing. We have come to see that if life is information, we cannot merely accept the hand evolution handed us. We could throw out the cards, mix up the deck, and print a brand-new game from scratch.

The globe watched in 2010 as the barrier was breached for the first time in a manner evocative of science fiction. Craig Venter, the same person who competed with the government to map the human genome, announced that his team had developed Synthia, the first-ever synthetic life form. They went beyond simply altering a current bacterium. They created a genome on a computer, bought the chemicals A, C, G, and T from a mail-order store, and then actually printed a fresh chromosome from scratch. They then inserted the created software into a hollowed-out bacterial shell. The bacteria booted up when the device was switched on. Following the instructions written by people, it started to expand and divide precisely as the computer code specifies. For the first time in four billion years, a living being existed whose ancestor was a computer file rather than a biological parent. We had evolved from editors of the ancient world to creators of the new one.

While Synthia's birth demonstrated our creative ability, a more recent study revealed our shocking efficiency. Researchers at the University of Alberta wanted to find out in 2017 if they could make a virus that was no longer found in nature. They selected the Horsepox virus, a relative of smallpox that had vanished decades ago. From an internet database, they only had the Digital Sequence, the list of letters; they lacked a real sample. For roughly $100,000, they mailed the matching DNA fragments and then joined them in a lab. They had a working, live virus able to infect cells within six months. This showed that the threshold is immutable: we can never really wipe it out as long as we have a pathogen or creature's Digital Code. Turning the Script of Life into a Print-on-Demand service, we have the power to raise the dead and produce the invisible.

The "Architect's Threshold" represents the death of Biological Luck. For the entirety of human history, your health, your intelligence, and your lifespan were a gamble - a roll of the genetic dice at the moment of conception. You were "blessed" with good genes or "cursed" with bad ones. As we step over this threshold, the gamble is being replaced by Design. We have realized that the laws of nature are not moral commands; they are just technical constraints. To the Architect, a genetic disease is no longer a "tragedy" or a "test of faith" - it is simply a debugging problem. We have realized that the sacred blueprint is actually a Wiki; it is open for collaborative editing, and we are the ones holding the pen.

The era of the Code-Breaker is ending. The era of the Bio-Designer has begun.

II

The Second Genesis (The Current Revolution & Quantum Secrets)

Part II explores the "Now" - humanity's shift from observing to rewriting life via CRISPR and Synthetic Biology. It reveals life's "Secret Engine": Quantum Biology, where enzymes tunnel and cells maintain coherence. Mastering these forces lets the Architect "Defy the Reaper" through slowed aging and cryopreservation, ushering in the Synthetic Canvas - an era of intentional, engineered life beyond the natural world.

5

CRISPR and End of Luck

"The power to control our species' genetic future is awesome and terrifying. Deciding how to use it may be the biggest challenge we have ever faced." *—Jennifer Doudna, Co-discoverer of CRISPR-Cas9*

For nearly four billion years, a silent, invisible war has been raging on every square inch of the planet. It is a war between bacteria and the viruses that hunt them, known as Bacteriophages. This conflict is the most ancient and brutal arms race in the history of life. Young researcher Francisco Mojica discovered a biological prize hidden in plain sight inside the DNA of these salt-loving bacteria on the sun-baked salt marshes of Santa Pola, Spain, in 1993. He discovered odd, recurring DNA patterns: short sequences reading the same both forward and backward, separated by enigmatic spacers. CRISPR is what he called them (Clustered Regularly Interspaced Short Palindromic Repeats). The scientific world disregarded him at the time. They believed he had come across trash. But Mojica had found something far more significant:

he had found the Memory of a Species.

CRISPR's brilliance resided in the Spacers, the intervals between the repeats, not in the repeats themselves. Mojica understood that these spacers were perfect genetic matches for the viruses that had attacked the bacteria before; they weren't just random nonsense. The bacteria were practically photographing their foe. When a virus attacked and the bacterium survived, it would snip a piece of the virus's DNA and stitch it into its own genome. This created a Molecular Most Wanted List. If that same virus ever attacked again, the bacterium would recognize the face (the code) and use a special protein, Cas9, to hunt it down and shred it.

This revelation, as Walter Isaacson so skillfully shows in his magnum opus of the CRISPR revolution, The Code Breaker, changed our whole view of evolution. Isaacson observes that it was once believed only complicated creatures had adaptive immune systems capable of learning from experience. Mojica showed that even the most basic forms of life on Earth have a complex, adaptable defense mechanism.

Mojica was postulating in Spain when the first real-world evidence of CRISPR came from a Yogurt Factory rather than from a medical lab. Mid-2000s, Danisco food company researchers were losing millions of dollars as viruses were killing bacteria needed to manufacture yoghurt. Rodolphe Barrangou and Philippe Horvath observed that some bacterial populations were intrinsically resistant to viral epidemics. When they analyzed the DNA of these survivors, they discovered the DNA of viruses within the CRISPR spacers of the bacteria. They physically transferred a spacer from an immune bacterium into a weak one to support their idea. The susceptible bacteria suddenly got vaccinated against the virus. People first purposefully altered an immune system here. Every time you have a cup of yoghurt nowadays, there's a good possibility you're consuming the byproduct of a CRISPR-enhanced bacteria colony meant to live through a viral plague.

Though it almost had a far less memorable moniker, CRISPR is today among the most well-known acronyms in science. Before scientists decided on CRISPR, other researchers were calling these sequences SRSR (Short Regularly Spaced Repeats), DR (Direct Repeats), or LURs (Large Unnamed Repeats). The person who first saw how important they were, Francisco Mojica, apparently thought of the name CRISPR over a few beers with a friend. He wanted a name that was descriptive but also "crisp" and memorable. It is a rare case in science where the branding was as perfect as the discovery, suggesting a tool that could "crisply" cut through the chaos of the genetic code.

Humanity didn't "invent" gene editing. We simply found a sword that had been lying in the dirt for a billion years and figured out how to pick it up. Nature didn't design CRISPR for us to cure cancer or design babies; it designed it for a single-celled organism to survive a microscopic mugging. The "treasure" wasn't a gift from the gods; it was a byproduct of the brutal, unceasing struggle for survival. We are the "Last Prometheus" not because we created the fire, but because we were clever enough to steal it from the bacteria.

Doudna and Charpentier: The Nobel Duo

2011 was the year the hammer at last met the anvil in the universe of the Last Prometheus. Although Francisco Mojica had found the wealth in the salt marshes, two women on opposite sides of the world - Jennifer Doudna in California and Emmanuelle Charpentier in Europe - figured out how to turn that treasure into a weapon. They wished to capture the tools of the bacteria and use them to rewrite the fate of all life on Earth, not only to observe the bacteria battling viruses. Their teamwork was a clinic in Synchronicity. Doudna was the top-notch structural biologist at Berkeley, while Charpentier was the restless microbiology nomad traveling from lab to lab all around Europe. They found they were both gazing at the same puzzle from various angles when they met at a meeting in Puerto Rico. They set out to see whether combining their hardware and software would produce something the world had never seen: a Programmable DNA Editor.

Doudna and Charpentier's actual brilliance was in their capacity to simplify the complexity of nature. CRISPR in the original bacterial system cuts viral DNA by using a messy mix of two kinds of RNA molecules to find and cut it. It was cumbersome and difficult to operate. They discovered as they worked together: Why use two molecules when one will enough? They came up with a way to make a single, fake strand called sgRNA (Single Guide RNA) by joining those two molecules. They began to view CRISPR as an Engineering Platform rather than a holy biological process. They could instruct the Cas9 scissors to go to any place in any genome - human, plant, or animal - and make a cut with surgical precision just by altering the code on that one guide.

The most important discussion of 21st-century biology in history did not take place in a sophisticated laboratory. During a scientific meeting in 2011, it occurred on a stroll along the small, cobblestone streets of Old San Juan, Puerto Rico. Charpentier and Doudna slipped away from the conference to grab a coffee at a local cafe. As they walked, Charpentier proposed the idea that they should work together to solve the mystery of how the Cas9 protein actually finds its target. Doudna later recalled that she felt a shiver of excitement. By the time they finished their walk, the alliance that would win the Nobel Prize was born. It was a reminder that while the Last Prometheus uses supercomputers, the biggest breakthroughs still start with two humans, a shared idea, and a simple conversation.

The Nobel Prize is renowned for moving slowly. Scientists typically wait 30 or 40 years following their finding to get the call from Stockholm, as the committee waits to see if their work holds up over time. Doudna and Charpentier published their landmark paper on programmable CRISPR in 2012. They were awarded the Nobel Prize in Chemistry just eight years later, in 2020. Almost unheard of, this Nobel Prize shows the sheer, explosive potential of their finding. Labs everywhere were already using their Single Guide to edit flower colours and cure mouse diseases within months of their publication. It took less than 30 years for the globe to recognize that the game's rules had permanently changed.

The realization of the "Nobel Duo" is that DNA is DNA. Before this, we viewed different species as separate kingdoms with different rules. Doudna and Charpentier proved that a tool from a humble bacterium could function perfectly inside a human cell. They revealed the modular nature of life. We are all running on the same underlying code, and now we have a "Universal Translator" that allows us to edit that code across all borders. We no longer have to wait for the slow, natural process of evolution to give us a trait. We can now paste a trait from one life form into another as easily as we move a file from one folder to another.

The Guide and the Blade: How CRISPR Works

Understanding the power of CRISPR requires you to view biology as a Word Document rather than as life. Consider a three billion letter document. One A instead of a G in that document causes the whole program to crash and leads to a genetic disease if there is only one typo.

Our editing techniques before CRISPR were like tossing a bomb at the paper and hoping the explosion would, by chance, rearrange the letters into the appropriate word. CRISPR transformed that. This biological "Search and Replace tool" has a cursor exactly one molecule wide. Two main elements make up the system:

1. **The Guide RNA (gRNA):** This is the "GPS." It is a synthetic strand of genetic code that we program to match the specific "typo" we want to find. It floats through the nucleus of the cell, ignoring billions of letters until it finds the exact 20-letter sequence that matches its own.
2. **The Cas9 Protein:** This is the "Blade." It is a massive, claw-like enzyme that hitches a ride on the Guide RNA. Once the GPS finds its target, the Cas9 unzips the DNA and performs a double-stranded cut, physically breaking the "spine" of the gene.

You might wonder, if CRISPR is so good at cutting, why doesn't it just shred the bacterium's own DNA? The response is a little sequence known as the PAM (Protospacer Adjacent Motif).

Let's understand this in a very simple way. Imagine bacteria are like tiny superheroes that get attacked by viruses (bad guys who want to sneak inside and mess things up). The bacteria have a special defense system called CRISPR — it's like their memory of bad guys they've seen before. When a virus attacks, the bacteria save a small picture (a short piece of the virus's code) in their memory book. That picture is called a spacer. Later, if the same virus (or a very similar one) tries to attack again, the bacteria send out a scout called Cas9 (think of it as a pair of super scissors). The scout (Cas9) looks for a matching piece in the virus's DNA — that's the part that looks exactly like the picture (spacer) they saved. But here's the clever part: the bacteria don't want the scissors to accidentally cut their own DNA!... So nature added a secret password right next to the matching part in the virus. That password is called the PAM (short for Protospacer Adjacent Motif). It's just a few letters (like "NGG" — very short, usually 2–6 letters of DNA code).

- If the scout finds the matching picture (spacer) + the secret password (PAM) right next to it → snip! It cuts the virus.
- If it finds the matching picture (spacer) but no password (PAM) → it says "This might be our own stuff!" and walks away without cutting.

The virus almost always has this password (PAM) next to the matching part (Spacer), but the bacteria carefully made sure their own memory book never has that password next to their saved pictures. So the scissors never cut the bacteria by mistake.

Let's make this more simple:

- PAM = the secret password sticker that must be right next to the bad guy's photo, otherwise the bacterial scissors won't cut.

- Without the PAM sticker, CRISPR scissors just ignore it - even if everything else matches perfectly.

That's how bacteria stay safe while fighting viruses... and that's also why scientists have to look for a PAM when they use CRISPR to edit our genes!

The bacterium's own Most Wanted List doesn't have these PAM sequences, so the Cas9 knows to leave the home DNA alone. If we use CRISPR on people, we need to find these natural locks to know where we are allowed to cut. It serves as a reminder that even the smartest instrument in biology depends on a basic, binary Yes/No check.

People most often misunderstand about CRISPR is that the tool itself simply breaks the DNA; it does not really change it. The cell's own frenetic repair equipment handles the editing.

How we use CRISPR to turn off a bad gene?

DNA is like a very long instruction book written in letters (A, T, C, G). Sometimes the book gets torn in two — a clean double-strand break happens in the DNA chain. Cells have emergency repair systems to fix these breaks quickly so the cell can survive. One main fast repair is called NHEJ (Non-Homologous End Joining). NHEJ is quick but very sloppy — it just grabs the two broken DNA ends and glues them back together without checking carefully. Because it's in a rush, it often makes small mistakes:

- It deletes a few letters (small deletion).
- It adds a few extra random letters (small insertion).
- These mistakes are called indels (insertions + deletions).

Even 1–2 extra or missing letters can ruin the whole sentence (gene). This often causes a frameshift — the reading frame shifts, so all following instructions become nonsense. This usually creates an early stop signal (premature stop codon), making the gene completely broken and non-functional. The gene is "knocked out."

If we want to turn off a bad gene (for example, one that helps cancer cells grow uncontrollably), we use CRISPR-Cas9 like very precise molecular scissors. We guide it to cut the DNA exactly in the middle of that bad gene, creating a clean double-strand break. Then we let the cell do its normal emergency repair — mostly NHEJ. Because NHEJ is error-prone, in most cases it adds or deletes a few letters right at the cut site. This almost always breaks the gene's instructions (frameshift + early stop), so the bad gene stops working. The cell has accidentally silenced its own harmful gene.

We don't need to perfectly rewrite or edit the DNA ourselves. We just make one clean cut in the right spot - the cell's own rushed, mistake-filled repair (NHEJ) does the job of ruining the gene for us. This is the main way CRISPR creates gene knockouts in research, medicine, and therapies today.

The He Jiankui Incident: The Forbidden Threshold

Every revolution in the story of the Last Prometheus has an Icarus, a person who flies too close to the sun and falls, bringing shame to the flames along with them. November 2018 saw He Jiankui, a Shenzhen, China, as that man. While the rest of the scientific world was arguing whether we should ever alter the human germline—the genes passed down to coming generations—He Jiankui realized that the moment for discussion was past. He crossed the last forbidden line: he brought human embryos to life after altering their code.

The world discovered this not via a peer-reviewed publication but rather by a series of slick YouTube videos. He revealed the birth of twin girls, Lulu and Nana, whose DNA had been changed with CRISPR to provide HIV resistance. That was the biological version of a gunshot heard all across the globe.

You must know the Red Line of bioethics to appreciate why this set off a worldwide panic. Most CRISPR research is somatic; that is, changing cells in a live person to treat a condition. If that person passes away, the edit goes with them. However, He Jiankui did a Germline alteration. He changed every cell in the twins' bodies, even their eggs, by altering the embryo. This was the point the Promethean flame could have turned into a wildfire, as Walter Isaacson writes in his book The Code Breaker. Isaacson talks about how Jennifer Doudna and other early scientists were scared, not only because there was no openness, but also because these changes would now be permanently part of the human gene pool. We had moved from "repairing" humans to "redesigning" the species without a global vote.

The most sad aspect of the tragedy may have been He Jiankui's own view of his work. He considered himself to be a pioneer on par with the inventors of IVF; he did not view himself as a villain. He Jiankui actually hoped to be praised as a hero before he spoke at the Second International Summit on Human Genome Editing in Hong Kong. He thought he would be the first Chinese scientist to get a Nobel Prize for a medical discovery. But the Truth was of the matter. He ran into a wall of icy, scholarly anger rather than applause. For unlawful medical practice, his colleagues rebuked him, his university fired him, and a Chinese court finally sentenced him to three years in jail.

He discovered the hard way that the world does not forgive arrogance when you play with the Blueprint of Life.

He Jiankui aimed at CCR5, a gene coding for a doorway that HIV uses to enter the cells. He aimed to render the twins resistant to the virus by disrupting this gene. But nature seldom is so straightforward. Neuroscientists made a shocking coincidence shortly after the news came out: studies on mice had shown that deleting the CCR5 gene not only affected immunity but also markedly improved memory and the capacity of the brain to heal from injury. This prompted a scary question: had He Jiankui accidentally produced the first cognitively enhanced people in the world? While we may never know the true status of Lulu and Nana's IQ, the incident proved that in the complex web of DNA, you can never change just one thing. Every edit has an echo.

The intellectual realization of the He Jiankui affair is that The Threshold cannot be un-crossed. We now live in a world where gene-edited humans walk among us. They are young, and they are private, but they are a new branch on the human tree. He Jiankui proved that the technology is now so accessible that a single "rogue" Architect can change the genetic future of humanity in a private lab. The "Sacredness" of the human genome was exposed as a fragile social contract, one that has now been permanently broken. We are no longer a species that can edit itself; we are a species that has edited itself.

Gene Drives: Editing the Earth

We have talked about altering people up till now—a particular embryo or someone's blood cells. But unlike individual targeting, CRISPR's god-like use goes beyond the person and aims at the entire species. This is the Gene Drive.

In natural biology, inheritance is a coin toss. A child of a father carrying a particular gene has a 50% probability of inheriting it. This is called Mendelian inheritance. But a Gene Drive undermines the system. It guarantees that every generation has a particular trait replicated onto both chromosomes using CRISPR. This produces a selfish gene that virtually 100% probability distributes throughout a community.

For the Last Prometheus, the Gene Drive serves as the ultimate ecological steering wheel. It enables us to change the fundamental makeup of a species in the wild, not only one life. We are altering the rules for the whole environment, not only the players in the game.

Africa is now seeing the most aspirational and divisive use of this technology develop. Every year, malaria kills hundreds of thousands of people; most victims are young children under five. We have battled the illness with bed nets and chemicals for decades, but the mosquitoes keep changing. Target Malaria, a research group, is working on a gene drive to kill the Anopheles gambiae mosquito. They want to release mosquitoes with a gene that makes them sterile. Since it is a gene drive, this infertility feature would spread quickly across the natural population. The mosquito population in a specific region would crash in a few years, and malaria would just disappear because its transmission is gone.

To monitor their mobility, the crew carried out a historic safe release of non-drive genetically altered mosquitoes in Burkina Faso in 2019. This was the first occasion such technology reached African soil, therefore starting a high-stakes chess game between human engineering and planetary evolution.

Millions of Bio-Tech insects have already been released in

the United States, even if the African initiative is the most well-known. Millions of genetically modified Aedes aegypti mosquitoes (bearers of Zika and Dengue) have been released in the Florida Keys by a firm named Oxitec. The reality is that people sometimes call these Mule mosquitoes. Their design includes a self-limiting gene. The kids die before they reach adulthood when lab-grown males mate with wild females. In some test regions, this has lowered the mosquito population by upwards of 90%. The fact that we have gone from poisoning the environment with chemicals to precisely changing the environment is evidence of our progress. The species' natural reproductive drive is being used as a weapon against its survival.

The intellectual realization of the Gene Drive is that there is no more "Outside." For all of human history, "The Wild" was a place that existed independently of us—a place governed by its own ancient, unyielding laws. With the Gene Drive, "The Wild" becomes just another laboratory. If we can decide whether a species lives or dies, or what traits it carries, then nature is no longer an independent force; it is an extension of human will. This brings a terrifying moral weight: if we accidentally "break" an entire ecosystem while trying to fix it, there is no "Undo" button. We have moved from being guests on this planet to being its System Administrators.

Curing the Incurable: Victoria Gray

The final test of any instrument for the Last Prometheus is not its conceptual brilliance but rather its capacity to end human suffering. As the globe argued over the morality of designer babies and rebellious scientists, a subtle miracle was ready for Victoria Gray. Victoria had Sickle Cell Disease, a cruel hereditary condition brought on by a single-letter DNA typo that produces hemoglobin. This misprint makes red blood cells develop into a sickle form, which results in terrible pain, organ damage, and a shortened life span as they get trapped in blood vessels. Victoria was the first American in 2019 to have her genetic code altered using CRISPR. This was a permanent rewrite of her biological destiny, not a short-term fix.

The approach Victoria was cured with was a stroke of brilliance, using an evolutionary loophole. Every one of us carries a gene for Fetal Hemoglobin, a high-performance variant of the protein we use in the womb to extract oxygen from our mother's blood. A genetic switch known as BCL11A switches soon after birth, turning off the fetal version and activating the adult version. Scientists used CRISPR to destroy the switch rather than attempt to repair the damaged adult gene. They fooled Victoria's body into believing it was back in the womb by cutting the BCL11A gene in her bone marrow. Her body started creating fetal hemoglobin once more; this doesn't sickle. They fixed an old, healthy function by using a controlled pause. Before the treatment, Victoria Gray lived her life in a state of constant fear. A simple change in temperature or a minor stressor could trigger a "crisis," landing her in the hospital for weeks on high-dose narcotics. Once the cells modified by CRISPR were reintroduced into her body, the transformation was complete. Victoria characterized the experience in interviews as though a weight had been taken off. She was able to play with her kids for the first time in her life without worrying about a fall. She started referring to herself as a Superwoman, not because she had acquired a new ability, but rather because she had at last been given the normalcy we other humans take for granted.

Although Victoria's narrative is a success, it exposes the Open Truth of the biotechnological age: Being precise costs a lot of money. As of 2024, the FDA-approved CRISPR treatment for Sickle Cell Disease (brand name Casgevy) costs about $2.2 million per patient. This excludes the months-long hospital stay and taxing chemotherapy needed for the bone marrow transplant. Though we have discovered the Cure for the Incurable, we have also built a world where the depth of your insurance policy may determine your genetic fate. The End of Luck is now a luxury item.

The Body Is Like a Patchable Operating System. Biological destiny is only a software bug. For millennia, a child born with Sickle Cell was regarded as unlucky or cursed. Today, we see children as a system operating an outdated version of code.

Victoria Gray is the living proof that our bodies are not "finished products." We are works-in-progress. By bypassing the broken "Adult" code and reactivating the "Fetal" code, we proved that we don't have to be slaves to the way we were born. We can "patch" the hardware using its own internal history. The "Sacredness" of the blood has been replaced by the "Utility" of the edit.

The Designer Baby Dilemma

We have spent the past few chapters learning how to fix the flaws in the human machine as the Last Prometheus. But once you can fix a flaw, you always find out you can upgrade the hardware. This is the jump from Enhancement (pushing a person beyond the natural limits of the species) to Therapy, which returns someone to a normal state.

Designer Baby is now a technological inevitability rather than a 1990s science fiction cliché. If we can remove the gene for a cardiac problem, why not insert a gene for more oxygen capacity? If we can turn off a gene for early-onset Alzheimer's, why not change the genes linked to rapid-fire neural processing? We are going into an age when the cradle is becoming a Design Studio.

Usually, we first test the idea on our closest friends before we even consider improving a human. Researchers in Guangzhou, China, gave a rather visual illustration of Enhancement's appearance in 2015. With CRISPR, researchers altered Beagle dog embryos to remove a gene known as Myostatin. The natural brake for muscle development in your body is myostatin; it informs your muscles when to stop growing. The brakes are compromised without it. The outcome was two Beagles, Hercules and Tiangou, who had twice the muscular mass of an average dog. They were simply more; they were not sick. They were quicker, stronger, and more muscular than any Beagle that had evolved for 4 billion years. The results of this study showed that optimizing the code may be as easy as eliminating one regulatory stop command. Pre-implantation Genetic Diagnosis (PGD) is already being used to create a lite version of the Designer Baby era, even if full-scale CRISPR editing of human babies is now prohibited in most of the world.

Couples having IVF can have their five-day-old embryos tested now. To look for thousands of genetic markers, several cells are extracted and examined. This was first used to help people avoid fatal illnesses like Tay-Sachs. But Genomic Prediction, a firm, started giving Polygenic Risk Scores. This lets parents select an embryo depending on, among other things, its probability of being taller, of having a lower risk of obesity, or of even having a statistically higher educational attainment—a proxy for IQ. We are curating these infants but not yet altering them. We are treating the next generation as though they are a product line from which we pick the best model now available from a digital menu.

The worst thing about the Designer Baby dilemma isn't that we might make monsters, but that we might end up with a Permanent Biological Underclass.

Millions of dollars would be required for the health, intelligence, and longevity CRISP upgrades, as we saw with Victoria Gray's $2 million cure. Only the very wealthy will be able to pay them. For the first time, the difference between the rich and the poor won't only be about the cars they drive or the houses they own; it will be coded into their very DNA. We run the risk of dividing the human race into two species: the Gen-Rich (those who are optimized, healthy, and long-lived) and the Naturals (those who are still subject to the biological draw's fortune).

A child is transitioning from a "Gift" to a "Project." When a child is a "gift" (a product of nature's luck), parents accept them as they are, flaws and all. But when a child is a "project" (a product of genetic choice), the parents become Consumers. If a child is "designed" to be a world-class athlete but ends up preferring poetry, is the child a "failure of engineering"? By removing "luck" from the equation, we also remove the unconditional acceptance that defines the human experience. We are trading our "humanity" for "performance," and we are doing it one letter of code at a time.

Bio-Hacking: CRISPR in the Garage

For most of human history, tall walls, white coats, and billion-dollar gatekeepers kept the Code of Life safe. A PhD, a government grant, and a sterile facility costing more than a little island were requirements if you intended to modify a genome. The Last Prometheus, however, can transform luxuries into things. The front line of the genetic revolution today is in a Mississippi suburban garage and a California kitchen, not only in Harvard or Oxford. This is the realm of Bio-Hacking, the democratization of the biological operating system. It is fueled by a radical theory: every citizen has the Right to Read and the Right to Write if DNA is just code. Garage Architect is emerging as FedEx delivers the tools of creation to your doorstep for the cost of a used bicycle.

Former NASA scientist Josiah Zayner, who famously left the institutional world to start The ODIN, is the face of this movement. DIY CRISPR kits sold by his business let anybody at home genetically modify yeast or bacteria.

Walter Isaacson portrays Zayner as the punk rock rebel of the CRISPR era in his book The Code Breaker. Isaacson explains how Zayner sees the established scientific world as a stifling hierarchy that hoards knowledge. Zayner sold an idea when he offered a CRISPR kit for under $200, not only a product. He showed that the Master Key to the human race's future is no longer a state secret; rather, it is an open-source tool that anyone with a credit card and a curiosity about the Digital Underworld of the cell may use.

October 2017 saw the most viral and divisive event for the bio-hacking movement. Josiah Zayner pulled out a syringe during a live-streamed conversation in a packed room and injected himself with a CRISPR cocktail intended to knock out the Myostatin gene in his left forearm. He aimed to demonstrate that Human Genetic Engineering was a current actuality, not some future dream. Basically, he wanted to turn his arm into a super-limb by stimulating regional muscular development. While the experiment didn't actually result in a "Popeye-sized" arm (the concentration of the CRISPR solution was likely too low, and the body's immune system probably neutralized it), the stunt was a massive success in social engineering. It forced the FDA to issue a public warning against DIY gene therapy. It signaled to the world that the "User" had arrived, and they weren't going to wait for permission from the "Developer."

For bio-hackers, the glowing-in-the-dark feature is the Hello World. Proving a genetic change has worked is most easily done this way. One bio-hacker, however, went too far. Beyond only making shining beer (which The ODIN offers a kit for), bio-hackers have tried to design their own skin to fluoresce, utilizing GFP (Green Fluorescent Protein) from jellyfish. Zayner experimented with a fluorescent tattoo method, injecting GFP DNA into his own skin. Although he ultimately didn't shine like a lamp—since human skin cells are significantly more difficult to reprogram than basic bacteria—the effort underlined a significant bio-safety issue. If you can make yourself glow, you can also accidentally trigger an immune reaction or an unintended mutation that leads to cancer. Bio-hacking's Cool Factor usually crashes head-first into the Reality Factor of complicated biology.

The Agnostic awareness of the garage lab is that Information wants to be free, even if it is alive. Once upon a time, we felt the public was too immature for some skills. Information

Gatekeepers determined who got to play with the code of life. Bio-hacking shows that these gatekeepers are no longer useful. It is impossible for you to un-invent a $150 CRISPR kit. We have entered a period when biological accountability is a personal decision chosen by people rather than a group choice chosen by governments.

Everybody has gotten fire from the Last Prometheus. We're no longer questioning if people should have this authority; We have to live in a world now where they already do.

The intellectual realization of the garage lab is that Information wants to be free, even when it's alive. We used to believe that certain powers were too dangerous for the public. We had "Information Gatekeepers" who decided who got to play with the code of life. Bio-hacking reveals that these gatekeepers are now obsolete. You cannot "un-invent" a $150 CRISPR kit. We have moved into an era where biological responsibility is no longer a collective decision made by governments; it is a personal choice made by individuals. The Last Prometheus has handed the fire to everyone. We are no longer asking if people should have this power; we are now forced to live in a world where they already do.

The Ethics of the Pen

We are now at the most challenging part of the Precision Era. We carry surgical scissors, a GPS, and the map. But as we grasp the Pen of Life, we must face a problem science cannot solve: We should write what?

The narrative of a human life in the past was shaped by the cruelty of the surroundings, together with the unpredictability of nature. We are now those who sit at the desk. However, a pen is only as good as the hand that wields it, and history reveals that humans frequently end up with a nightmare when they aim to define perfection. This is the Ethics of the Pen: the awareness that someone else's mistake is another's characteristic of identity.

Physician and author Siddhartha Mukherjee gives a sobering warning about the force of the Architect in his magnum opus, "The Gene: An Intimate History." He argues that our definitions of "normal" are not fixed biological truths, but mirrors of our own current prejudices. Mukherjee says that defining beauty only as blue eyes will help us find a gene for beauty and get rid of everything else. If the history of the last century taught us the dangers of empowering governments to determine genetic 'fitness'... the question that confronts our current era is what happens when this power devolves to the individual. He believes that we can inadvertently erase the very weakness that defines us as human, sensitive, and creative if we attempt to wipe away infirmities.

Recombinant DNA existed before CRISPR. Scientists first discovered in the early 1970s how to transfer DNA from one organism to another. The scientific community did something remarkable rather than hurrying forward: They cut off. 140 attorneys, ethicists, and scientists met at the Asilomar Conference in California in 1975. Until they could show their work was safe, they announced a self-imposed halt in it. They developed the initial bio-containment guidelines to guarantee that man-made bacteria couldn't leave the lab. This was one of the few times the Prometheans tried to control their own fire. Critics contend, though, that a Modern Asilomar could be impracticable given the current fast speed and highly decentralized nature of the current CRISPR revolution (see:

Bio-Hacking). The Pen is in too many hands for everyone to concur on one narrative.

The Deaf Community presents among the most serious ethical conflicts in the CRISPR age. Many doctors view deafness as a treatable illness. For many in the Deaf community, it is a culture and a common identity. Certain genes, like GJB2, lead to genetic deafness. CRISPR could theoretically eradicate deafness permanently from the human gene pool. Many others contend that this is a kind of genetic cleaning. They emphasize that a society without Deaf genes loses a vivid community, a distinctive approach to seeing the world, and a special language (Sign Language).

This forces us to ask: Is a "disability" a biological mistake, or is it just a biological variation that society has failed to accommodate? If we use the Pen to delete every variation we find "inconvenient," we may end up with a world that is "perfectly" boring and dangerously fragile. But as far as I'm concerned, we shouldn't accept any kind of disability when we have the power to control it. Just think for q while, your wife conceives an embryo, and you come to know that your baby is gonna blind or deaf for his whole life. Ask yourself, will you let that happen to that innocent life? Will you let the cruel nature continue making mistakes? Will you compromise with that and justify the mistake of nature? What will your kid think of you when he or she comes to know that you had the power to make the corrections, but you chose not to?

If someone thinks that our corrective measures will wipe out a specific language (sign language), then ask them to rupture their own eardrum and keep that language alive. We're not gonna let our generation suffer for nothing.

The End of Luck

We have reached the biological frontier's verge. We followed the route of the Last Prometheus in the chapters before, from the unintentional discovery of bacterial immunity to the surgical accuracy of the human operating system. We are no longer living in a world of theoretical possibilities as of 2026; This is the first decade of the Post-Luck Era.

Throughout all of human history, biological identity was a lottery. You played your cards—some high, some low—till the game came to an end. It was thought to be a calamity, a stroke of terrible luck, or the will of a higher authority if you were born with a bad gene. That narrative has fallen apart today as CRISPR therapies like Casgevy (It works by editing the patient's own blood stem cells to increase fetal hemoglobin, reducing painful crises and eliminating transfusion needs) become routine clinical treatments for blood disorders and clinical studies for everyday illnesses like high cholesterol reveal nearly 60% LDL decreases from a single injection.

Psychological, not technological, is the most major change in the End of Luck. A parent of a child with a genetic illness questioned in the past, "Why me?"—a query directed to the cosmos. That issue in the CRISPR era changes to, Why wasn't the code fixed? Once a defect becomes editable, it stops being a fate and starts being a choice. This generates a fresh and heavy load of Genetic Accountability. We are equally accountable for the result if we have the Pen of Life and decide not to use it as we would have been had we used it and failed. The One-and-Done cure is replacing a lifetime of Luck with a single, permanent engineering choice as the CRISPR market is forecast to reach almost $9 billion by 2027.

In early 2025, preliminary results from the Heart-2 clinical trials (using the VERVE-102 base editor) sent shockwaves through the medical community. Patients with life-threatening hereditary high cholesterol received a single infusion of CRISPR-based editors. Unlike traditional drugs that you take every day, this editor traveled to the liver and permanently "flipped a switch" to lower cholesterol production.

Data from 2026 indicated that, even without a single tablet or statin, these patients' cholesterol levels stayed at record lows for more than a year. It signaled the point at which we quit treating a chronic condition and began writing it out of reality. We altered their biological baseline in addition to their hearts.

We have perfected the "How" of editing, but the "Where" is still a patchwork of worldwide law. More than 75 nations have so far adopted international standards or national laws expressly prohibiting human germline (heritable) editing as of 2026. We have erected a legal barrier to keep us from rewriting the future of the whole species too rapidly even as we rush to cure the living. Though we have the technical ability to influence our development, at least for the time being the wisdom of the group maintains the Save button locked.

The end of luck is the beginning of total responsibility. When we were subjects of nature, we could blame our flaws on the stars or our ancestors. But as the architects of life, we have no one left to blame. If our children are born with bugs in their code, it is because we allowed them to be there. If we create a genetic wealth gap, it is because we designed it.

We have successfully stolen the fire, and in its light, we can see every letter of our own code. The fire has burned away the "Luck" that used to protect us from our own choices. We are now the authors of our own biological story—and the pen never stops writing.

6

God's Pencil - Toolkit of Biotechnology

"The rewards for biotechnology are tremendous—to solve disease, eliminate poverty, age gracefully. It sounds so much cooler than Facebook." — *George M. Church*

The New Age of Biotechnology begins like the first light breaking across a long, dark horizon. Imagine a world where the old scourge of organ failure no longer claims fatalities with terrible inescapability. Early in 2026, over 100,000 Americans alone are waiting on the national transplant waiting list—over 103,000 according to the most recent statistics from organ-donor.gov and UNOS, with some estimates reaching 121,000, including more general data. With around 90,000–100,000 people waiting and around 13 people dying daily before a suitable organ arrives, kidneys lead the desperate count. The situation resonates much more strongly all around. Still, biotechnology is quietly revolutionizing this shortage in labs and hospitals all around the world.

Imagine a youngster born with a rare genetic condition

once destined to live a short, painful life; now, innovative treatments rewrite cellular instructions to return function. Or think about the farmer watching his crops dry up in the heat and not getting enough water. Synthetic biology engineers can make microbes that capture carbon better, make biofuels that are better than what we have now, or make crops grow faster so that we can feed billions of people and cut emissions. One recent study estimates that synthetic biology might sequester up to 30 billion tons of CO_2 — the lifetime emissions of every passenger car ever built, by improving crops, lowering livestock methane, and using precision fermentation for low-impact foods.

These aren't far-off illusions. Record numbers of organ transplants (over 49,000 in the U.S. in 2025) indicate advancement in 2025–2026, but the gap grows as demand exceeds supply. Elegant, strong instruments from biotechnology close that gap, basic in theory yet extremely significant in execution. The methods discussed in this chapter point to a golden era of creativity. They enable us to clearly read the code of biology, make exact corrections or additions, and alter living systems from individual molecules to whole ecosystems. We are evolving to co-authors rather than passive spectators of nature's creation.

These techniques match in aim and curiosity but omit the headline-dominating CRISPR. They preserve life in suspended animation, heal our own cells to generate tissues, print layer-by-layer live structures, teach our bodies to create medicines from within, design animal organs for human use, utilize artificial intelligence to decipher protein mysteries, and design biology from first principles. Ethically, realistically, and with ever-increasing speed, they together create a toolkit that heals the incurable, keeps the planet warming, and pushes the bounds of human potential.

Cryopreservation: Freezing Life for the Future

Fade in on a clean chamber lit in delicate blue light. At -196°C, Dewar flasks are like silver sentinels holding liquid nitrogen. They are very tall. Inside cells, tissues, embryos, or in the most extreme examples, whole human corpses kept at the border of death are hanging in glassy stillness. Cryopreservation is the skill and science of stopping biological time without damaging the fragile structure of life.

The process starts with thorough preparation. Water inside cells changes to ice when temperatures fall. This makes the cells expand and break membranes like sharp glass. Scientists introduce cryoprotectants, particular chemicals like glycerol, dimethyl sulfoxide, or sophisticated vitrification cocktails, that replace water and encourage a glass-like condition known as vitrification to help avoid this catastrophe. No crystals develop; rather, the substance hardens into a non-harmful amorphous solid. Until the sample achieves cryogenic stasis, cooling ramps down progressively—usually assisted by immersion in liquid nitrogen vapour or controlled-rate freezers.

The wonder lies in reversibility's promise. What freezes can, in theory, thaw—if we master the return journey. Rewarming must be rapid and uniform (nanowarming with magnetic nanoparticles or radio-frequency fields shows promise in recent studies) to avoid devitrification cracks or toxic cryoprotectant buildup. When successful, cells resume function as if no time has passed.

There are already plenty of actual successes. One of the most proven successes of fertility preservation is oocyte (egg) or embryo cryopreservation by vitrification, which usually results in survival rates of over 90–95% following thawing. In several studies, cumulative live birth rates from frozen oocytes range from 30–35%; thousands of healthy babies have been born worldwide from eggs frozen for years or decades. This gives great power to people with cancer or those putting off having children; it's like a biological pause button that works against age or treatment.

Organ banking is becoming real. Among the advances of 2025–2026 are Until Labs' $58 million raise to expand reversible organ cryopreservation, Texas A&M's crack-prevention techniques for subzero storage, and rat kidney trials at the University of Minnesota showing almost perfect function after vitrification and rewarming. Companies aim at indefinite donor organ storage, therefore turning the present hours-long viability window into days, weeks, or longer—perhaps eliminating waste (tens of thousands of organs thrown away yearly) and facilitating worldwide matching.

Next is the frontier: whole-body or neuro-preservation cryonics. With costs ranging from $28,000 to $220,000, organizations such as Alcor (over 200 patients cryopreserved as of 2026) and others all around provide this to terminal patients, with the total surpassing 500. Though no person has been properly rewarmed, some see it as a link to future nanomedicine or repair technologies, as it is still theoretical. Ethical arguments rage, but the fact that the process keeps going shows that people don't want to believe things are over.

Cryopreservation helps conservation outside of people: frozen sperm, eggs, and embryos from threatened species help build frozen zoos for comeback plans. It may allow long-duration missions in space exploration: astronauts in induced torpor or seeds/embryos kept for far-off colonies.

Cryopreservation is not just about storing things; it's about fighting entropy and making a bet on what science will be like in the future. It whispers that once halted, life may one day press play—giving organs to the waiting, children to the hopeful, species to the endangered, and possibly even second chances to those who dare to dream of resurrection. The future in this modern age of biotechnology is not set in stone; rather, it is waiting, possible, and very much alive.

Induced Pluripotent Stem Cells (iPSCs): Turning Back the Clock on Cells

This story about a Kyoto lab in 2006 essentially discovered a reset button for life; therefore, it is not only about biology. Shinya Yamanaka took a normal, boring mouse skin cell and did something that sounded like science fiction: he turned back its biological clock.

He observed that a particular adult cell opened its identity to transform into a pluripotent stem cell by inserting four certain genes (Oct4, Sox2, Klf4, and c-Myc), today widely recognized as the Yamanaka Factors. This implies a cell capable of transformation into anything from a beating heart cell, a firing neuron, or a section of a lung. Six years later, he received a Nobel Prize for this discovery.

Though simple, the process is sophisticated. It starts with a basic sample—often urine, blood, skin or even cells. Though most of it is locked away as we get older, these cells already contain your full genetic blueprint. Scientists reactivate embryonic programs that have been dormant for decades by using reprogramming factors (delivered via viral vectors, non-integrating mRNA, Sendai virus, or increasingly efficient chemical cocktails).

Once these cells settle into induced Pluripotent Stem Cells (iPSCs), they may be directed into the three main lineages that make up a human being:

- **Ectoderm**: Making your skin, nerves, and brain.
- **Mesoderm**: Develop your bones, muscles, and heart.
- **Endoderm**: Making your inside organs, like the lungs and liver.

The real magic happens in directed differentiation. By mimicking embryonic signaling pathways, scientists use growth factors like Activin A for endoderm, BMP4 and FGF for mesoderm, or retinoic acid and sonic hedgehog for neural lineages to guide iPSCs into specific cell types. Current protocols create high-purity populations. These include cardiomyocytes beating in dishes, insulin-producing beta cells responding to glucose, retinal pigment epithelium (RPE) sheets prepared for transplantation, and dopaminergic neurons firing action potentials.

Stem cell research was mired in ethical issues before iPSCs because it needed embryonic tissue. Bypassing that wholly are iPSCs. These cells contain your identical DNA since they originate from you. This has two really big benefits:

- **No Rejection:** Your immune system recognizes the new tissue as self, which can eliminate the requirement for immunosuppressive medications taken for life.
- **Disease in a Dish:** Doctors could grow a patient with

Parkinson's own neurons in a petri dish. Before the patient even sees a pill, they can monitor the disease progression in real-time and try thousands of medicines on those particular cells to find out what works.

We are now mending people with these cells rather than only studying them.

- **Reversing Diabetes:** Chinese scientists reached a major turning point in 2024. For a Type 1 diabetic patient, they made working, insulin-producing cells using chemically induced iPSCs. The outcome? The patient returned to insulin independence. It was a practical reversal, not only a treatment.
- **Renewing Vision:** For iPSC therapy, the eye is among the first frontiers. Scientists are currently producing sheets of retinal cells for those suffering from age-related macular degeneration (the main cause of blindness). Trials like those from BlueRock Therapeutics, which have progressed into later phases and have really aided persons in recovering usable eyesight, have advanced since 2025.
- **Brain and Spine Healing:** Perhaps the most amazing use here is this one. To restore movement, Parkinson's patients' brains are now being tested for iPSC-derived dopamine neurons. Likewise, scientists are trying to grow new neural connections to get around the damage from spinal cord injuries.

Looking ahead, iPSCs will evolve in ways we can hardly imagine. By the 2030s, we can expect seamless integration. AI will design custom differentiation paths in hours instead of months.

Universal iPSC banks, modified through gene edits, will provide "ready-to-use" tissues around the world. Organoids, which are miniature 3D hearts, brains, and kidneys made from iPSCs, will develop into vascularized, transplantable mini-organs, helping to ease donor shortages.

Regenerative therapies could reverse the signs of aging by rejuvenating old tissues and resetting epigenetic clocks, potentially extending how long we stay healthy. When combined with synthetic biology, we could engineer iPSCs to produce needed treatments inside the body or create chimeric organs that mix human and animal cells for xenotransplants.

In a remarkable turnaround, imagine personalized longevity clinics where your skin cells can become a young heart valve, a neural graft, or pancreatic islets. These would be tailored, printed, and implanted. Conditions like blindness would be cured, paralysis reversed, and diabetes eliminated—not as miracles, but as standard practice. iPSCs don't just turn back the cellular clock; they shift humanity's biological future, leading us into an era where "incurable" is a thing of the past.

Organoids: Mini-Organs Grown in the Lab

Imagine looking through a microscope at a tiny, translucent sphere the size of a grain of rice. This sphere is alive and self-organizing into layered structures that mimic a full human organ. Welcome to the world of organoids, which are three-dimensional, miniature organ models grown entirely in the lab from stem cells. A group of pluripotent stem cells, often induced pluripotent stem cells (iPSCs) taken from a patient's own skin or blood, can change, under specific chemical signals, into a functioning version of brain, kidney, liver, intestine, or lung tissue.

The process resembles embryonic development in miniature. Stem cells are placed in a supportive matrix, typically Matrigel or synthetic hydrogels, and exposed to growth factors that imitate the signaling pathways of early organ development. For brain organoids, factors like FGF2 and EGF promote neural development. For kidney organoids, Wnt activators and retinoic acid help form nephrons. Over weeks to months, cells self-assemble: progenitor cells turn into specialized types—neurons and glia in brain models, hepatocytes and structures like bile ducts in liver organoids—creating complex architectures complete with lumens, polarized layers, and even early vascular-like networks in more advanced versions.

The amazing part is their striking resemblance to real organs. Unlike flat 2D cell cultures that lose context and function, organoids maintain spatial organization, cellular diversity, and organ-level activity. Brain organoids develop layered regions similar to a cortex, complete with working synapses and electrical signals. Intestinal organoids create structures that absorb nutrients and secrete mucus, while liver organoids metabolize drugs using cytochrome P450 enzymes. By 2025–2026, multi-organoid chips are expected to combine up to nine tissue types (liver, kidney, heart, lung, brain) under treatment, simulating how different organs communicate, like the gut-liver axis or blood-brain barrier effects.

This accuracy changes the field of biomedical research dramatically. Personalized drug testing becomes possible. Organoids made from patients help test treatments based on individual genetics, predicting effectiveness and potential side effects before clinical trials. In cancer research, patient-derived tumor organoids (PDTOs) model diverse cancers, allowing for fast, customized drug responses. Biobanks for sarcoma now capture subtype diversity for targeted treatments. Research on infectious diseases benefits as well: studies on the Zika virus's effects on brain organoids revealed mechanisms of microcephaly, while effects from SARS-CoV-2 on lung and brain models guided early treatment strategies.

Our understanding of diseases has advanced significantly. Brain organoids from patients with Alzheimer's show amyloid-beta plaques, tau tangles, and network problems, enhancing our understanding of neurodegeneration. Kidney organoids can model polycystic disease or drug toxicity. Liver organoids offer better predictions of drug-induced liver injury (DILI) compared to animal models. The FDA's guidance in April 2025 to reduce some animal testing in favor of organoids and organ-on-a-chip systems signifies a major regulatory shift and boosts adoption.

A significant achievement is the reduction of animal testing. Organoids provide human-relevant platforms that capture species-specific reactions that animal models often overlook, improving both translation and ethics. The global market for human organoids, valued at about $1.1–1.3 billion in 2024–2025, is expected to grow to between $2.7 and $13 billion by 2030, driven by these applications.

Yet, organoids also serve as a pathway to creating lab-grown transplant organs. Vascularized kidney or liver organoids are getting closer to functioning properly; when combined with bioprinting or microfluidics, they may lead to large-scale, rejection-free tissues. Challenges remain, such as limited size (nutrient diffusion issues), incomplete maturation, and reproducibility, but advances in bioreactor perfusion, AI-guided processes, and integrating multiple omics are expanding possibilities.

In this exciting development, organoids are more than just models; they connect laboratory work with patient care, unraveling human biology on a small scale. They signal a future where diseases are researched, drugs are customized, and organs may someday be regenerated in vitro.

3D Bioprinting and Tissue Engineering: Printing Living Structures

The scene is set in a laboratory, and a robotic arm is moving around, putting down ink in a very precise way. This live ink is made up of cells, biomaterials and things that help them grow. This process is called 3D bioprinting. It is a way of making things by adding layers. It is used to make models of tissues and organs that are guided by computers.

3D bioprinting is better than the usual way of making tissues because it gives us more control. The usual way uses a framework with cells stuck to it. This often does not work very well because the cells do not always end up in the right place. With 3D bioprinting, cells are put where they are needed, just like they would be, in a real body.

The basic methods of printing are different. Printing with extrusion is the technique that people use the most. This method pushes hydrogels that have cells in them through nozzles. Some methods use lasers to make layers solid by shining light on them. The inkjet method is very precise. It releases a very small amount of liquid, just the right amount. Bioinks are like helpers; they are hydrogels such as gelatin methacryloyl, which is also called GelMA or collagen or alginate or decellularized extracellular matrix, which is also called dECM. These bioinks have to be good for printing, and they have to keep the cells alive, usually than 90% of the cells have to be alive, and they have to be safe for the body. After the printing is done, the structures need to be stabilized. This is done by using ultraviolet light or special ionic techniques or enzymatic processes, which is like using special helpers to make the structures strong.

One big advantage of this technology is that it can be customized. We can use patient-derived cells to make constructs that reduce the risk of rejection. For burn victims, bioprinted skin integrates dermis and epidermis with vascular channels, accelerating healing and reducing scarring—clinical prototypes already show promise.

Bioprinting can also be used to repair cartilage. For example, we can make cartilage that matches the shape of a person's ear or joint. This can be really helpful for people who have osteoarthritis or who were born with defects in their joints.

Recently, scientists have made some breakthroughs in making blood vessels. They have figured out how to print networks of vessels using special cells called endothelial cells. These vessels can even carry blood, which is a really important step forward. This helps with the issues that diffusion has when it comes to tissues. These tissues can be really thicker than 200 to 300 micrometers.

The things they have planned for 2025 to 2026 are really something. Carnegie Mellon has come up with this method called FRESH (Freeform Reversible Embedding of Suspended Hydrogels). It is a way to make structures that can be used in the body. These structures are like homes for cells, and they are made out of collagen. They are called CHIPS.

The people at Carnegie Mellon are using CHIPS to try and

find treatments for Type 1 diabetes. They are putting these CHIPS into machines called bioreactors. This helps to create tissues that are, like the ones found in the pancreas. These tissues can even make insulin when they detect glucose.

They are using a kind of printing to make all this happen. This printing can use different materials. It helps to put the cells and proteins in the right places. This is important because it creates the environment for the cells to grow and work properly. Carnegie Mellon's FRESH method and these CHIPS are really interesting. The company is adding features to its products using a special kind of printing called cryogenic printing. They are also using something called 4D elements. These 4D elements are really cool because they can flex and change when you do something to them, like give them some input. The cryogenic printing and 4D elements are what make these new features possible.

The uses of this technology cover a range. Doctors are starting trials for things like burn skin grafts and reconstructive surgery. This technology helps with regeneration by using muscular or bone patches. We can also use it to grow meat in a lab, like beef or chicken, which helps to lower the impact of raising animals. This is a sustainable way to produce food.

The technology is also used for disease modelling and drug testing. Scientists make prototypes of organs like the liver, the heart or the kidney. The vascularization of these organ prototypes enables them to be more viable and useful for testing.

Still, obstacles exist. Among these are scalability, immune integration, complete vascularization for solid organs, and long-term functioning preservation. Progress is shown by discoveries such as pre-vascularized patches and robots capable of in-body printing. 3D bioprinting is moving tissue engineering from an art form to precision engineering. Regenerative medicine is moving towards standard health care practices as it is producing custom body parts that can heal, replace, and maybe even replicate organs.

mRNA Therapeutics: Turning Cells into Drug Factories

Imagine a kind of syringe. It does not have the kind of medicine inside. Instead, it has some specific instructions that are biological. These instructions are called messenger RNA or mRNA for short. When you put this into your body, the mRNA goes into your cells. It is like a plan that your cells can follow. The parts of your cells that make proteins, called ribosomes, read this plan. They use it to make the proteins that your body needs to get better. After a day, the mRNA is gone. It breaks down by itself. Does not change your body's cells in any permanent way. This is what we call mRNA therapeutics. It is, like a way to turn your body's cells into factories that make proteins that can help you heal. The mRNA therapeutics do this by giving your cells the instructions they need to make these proteins. This is a change from putting proteins directly into something to letting cells make proteins quickly and, in a good way. Cells can now produce proteins on their own. This happens quickly and effectively, which is a great thing for cells to make proteins.

The COVID-19 pandemic made people notice mRNA because of vaccines like Pfizer-BioNTech's BNT162b2 and Moderna's mRNA-1273. These COVID-19 vaccines saved a lot of lives by making people's immune systems very strong. mRNA can do a lot more than just prevent COVID-19. mRNA is special because it can make proteins quickly inside the body in tissues or all over the body without using viruses or changing the body's genes forever. The mRNA is protected by particles called lipid nanoparticles, which help the mRNA get into cells easily. Chemical changes, like using pseudouridine, can really help with reducing the response. At the time, pseudouridine enhances the translation of the chemical changes.

Self-amplifying mRNA, which is also known as circular RNA variants, is pretty cool. These things can help with expression at lower doses of the chemical changes, and they also improve the stability of the changes.

This platform is really good for medicine, and that is because it is very flexible. When it comes to treating cancer, the platform uses something called mRNA to help the body fight the disease. The mRNA is special because it tells the system to go after the bad cells in the tumor. These bad cells have glitches that are not found in healthy cells, and the mRNA helps the immune system find them. For example, Moderna's mRNA-4157, which they made with Merck, works with another medicine called pembrolizumab, also known as Keytruda. Together, they help the immune system target the problems in the tumor, which makes the treatment more effective for each person. The results of Phase 2 showed that people with high-risk melanoma had a 44% risk of the cancer coming back. After five years, the risk of the cancer coming back or causing death was still 49% lower. We are waiting to see the results of Phase 3 by 2026. These results are for melanoma, non-small cell lung cancer, bladder cancer and renal cell carcinoma. If everything goes well, this could be the time that mRNA cancer vaccines are approved. There are studies going on to see if we can use a different way to help the immune system fight cancer. This is done by putting the treatment into the tumor. The goal is to activate the system to fight the solid tumors.

Some people are born with diseases that are caused by problems with proteins in their bodies. When these proteins are missing or do not work properly, it can cause problems. But now there is a way to help these people, it is called protein replacement. This is where a special molecule called mRNA helps the cells in the body make the missing protein. This way, the body can get the protein it needs without having to make changes to its genes.

Moderna is working on a treatment for a disease called propionic acidemia. Propionic acidemia (PA) is a rare, inherited metabolic disorder caused by a deficiency in the enzyme propionyl-CoA carboxylase, which breaks down certain proteins and fats. This leads to a toxic buildup of organic acids in the blood and urine, causing severe early-onset symptoms like poor feeding, vomiting, lethargy, and hypotension. They are using something called mRNA-3927 for its treatment. They are also working on a treatment for methylmalonic acidemia, this one is called mRNA-3705. Both of these diseases are caused by problems with the way the body breaks down food, which can cause toxic acids to build up. Moderna is trying to find a way to help the liver work so it can break down food and stop the disease from getting worse.

They are also working on a treatment for fibrosis, a disease that is caused by a problem with a protein called CFTR. The goal is to help the body make this protein so it can work properly. People with cystic fibrosis can feel better. In hemophilia, the mRNA is really helpful because it can help our body make clotting factors VIII or IX for a while. This can really cut down on the number of bleeding episodes that people with hemophilia have to deal with. The mRNA works with other methods, like siRNA, and one example of this is fitusiran. People think that fitusiran will be approved in 2025 to help people with hemophilia A and hemophilia B.

When we are dealing with pandemics, it is really important that we respond quickly. The good thing about mRNA is that it is very flexible, so we can make changes to it fast to deal with variants or pathogens. Moderna is working on a lot of things, like a flu vaccine, which is called mRNA-1010 and a vaccine for RSV (Respiratory Syncytial Virus), which is called mResvia and has already been approved. They are also working on a vaccine for mpox. The plan is to have options for seasonal vaccines in the future. Moderna's pipeline is really important because it includes all these vaccine candidates, such as the flu vaccine, the RSV vaccine and the mpox vaccine.

There are problems that we need to deal with when it comes to mRNA therapeutics. For example, we need to improve how we deliver these therapeutics to the place in the body. A lot of the time, mRNA therapeutics end up in the liver. We are getting better at making them go to specific tissues. We are also working on making sure that these therapeutics do not just work for a time.

The thing is, our body's immune system can react to mRNA therapeutics, and that is a concern.

People are working on ideas to make mRNA therapeutics better, like using computers to design special sequences and attaching antibodies to help with delivery.

mRNA therapeutics are changing the way we approach treatment: cells are becoming programmable factories that deliver specific proteins as needed. From personalized cancer immunotherapies to rapid responses for infectious diseases and relief for rare conditions, this technology promises a future where medicine can be as quick as sequencing a genome and as targeted as a patient's individual biology.

Xenotransplantation: Animal Organs for Humans

The operating room is really quiet. You can feel how focused everyone is. Machines are keeping an eye on the patient's signs as the doctors get ready to connect the blood vessels. They have an organ from a pig that has been carefully changed at the gene level. This organ is kept cool. It is still good to use, and it is really similar to a human organ. This is what they call xenotransplantation. It is a way to use organs from one species to get as many organs as they need, so people do not have to wait for a long time to get a transplant. The waiting lists are a problem because people lose hope every day. Xenotransplantation is a way to solve this problem by using organs from animals, like pigs.

A lot of Americans are waiting for a transplant. As of 2026, there are more than 100,000 Americans on the list. The number of people waiting is around 103,000 to 121,000. This depends on who you ask. Most of the people on the list need a kidney. Over 90,000 people are waiting for a kidney.

This is not a problem in the United States. Around the world, millions of people are waiting for a transplant. Sadly, thousands of people die every year because they cannot find a donor.

Most transplants happen because someone donates an organ. In 2025, there were 49,000 transplants in the United States. The problem is that there are no organs being donated. At the time, more and more people need transplants because they are getting older or have chronic diseases. The number of people who need transplants is going up. The number of organs being donated is not.

Pigs are really donors because their organ size is similar to humans. The way their bodies work is also similar to ours. For example, the rate at which their kidneys filter waste is a match. We can also breed a lot of pigs quickly.

The problem with using pigs as donors is that our bodies might reject their organs. This is because we already have antibodies that fight against a type of sugar called alpha-gal that is found on pig cells. There are also issues with blood clotting and the risk of getting viruses from the pigs. Pigs have this sugar called alpha-gal that can cause problems when we try to use their organs in humans.

The thing about CRISPR-enabled gene editing is that it gets rid of a lot of problems. Companies like United Therapeutics and eGenesis are working on something called "source pigs". United Therapeutics has something called UKidney. It has ten edits. Six of these edits are genes. These human genes are things like complement regulators CD46, CD55, CD59, thrombomodulin and endothelial protein C receptor. They help prevent the body from rejecting the pig kidney. They stop clotting. The other four edits are porcine genes that they get rid of. These are things like GGTA1 for alpha-gal CMAH, β4GalNT2 for glycans and growth hormone receptor. They do this to control the size of the pig.

EGenesis is going further with this. They are making sixty-nine edits. They are getting rid of endogenous retroviruses, which are called PERVs for short. They are also fixing other problems that make it hard for humans and pigs to be compatible. CRISPR-enabled gene editing is really important for this. CRISPR-enabled gene editing is what makes all of this possible.

Big things are happening with transplants. The way was made clear by cases where doctors helped people who really needed it. For example, NYU Langone did kidney transplants using pig kidneys in people who were still alive from 2024 to 2025. One of these people lived for a month before having some problems.

In 2025, the FDA said it was okay to do some important tests. Then, United Therapeutics started a study called EXPAND. This study began in November 2025 at NYU Langone. They did the transplant of a pig kidney, also called a xenokidney, into a person. This person had bad kidney disease and could not get a human kidney. This study starts with six patients. Then, more patients are added, so there are 50 patients in total. The doctors watch these patients for 24 weeks after they get a transplant. Then they keep watching them for the rest of their lives. They want to see how well the transplant works, and if the patients get any diseases from animals. The first results are good because they are similar to what happened in tests with baboons. The transplant seems to be working in these animals, so the doctors think it might work well in people, too, and the patients might live longer.

eGenesis got approval from the FDA for a trial. This trial is for a kind of kidney that has been edited 69 times. The goal of this trial is to see if it is safe and works well for people. The company is planning to test this kidney on dozens of people over the next few years. This will happen in different stages.

The people at eGenesis are already doing some of these procedures in 2026. Some of the patients who got these kidneys are doing okay. They have been living with these xenokidneys for a few months. They are working as they should. There are still some problems that can happen. Sometimes patients get infections, have trouble with the blood vessels. eGenesis is working on these eGenesis kidney trials to make them better. The eGenesis kidney trial is very important for the people who need a kidney.

These new applications are really promising. They can change things a lot. For example, they can make kidneys that work, so people do not have to go on dialysis. This can help a lot of people in America than the 90,000 people who are waiting for a kidney. It can also save a lot of lives every year.

The same thing can be done with hearts, livers and lungs. There is some data from tests and some cases where people have gotten these organs, like a pig heart, and it has worked out.

There is also something called bridging transplants. This is when doctors use xenografts, which are organs from animals, to help people until they can get a human organ.

The way we think about ethics and safety is changing. We have to make sure we are screening for germs really carefully. We also have to get permission from people and treat animals well. Luckily, we do not see a lot of animals giving diseases to people. When we do tests, we watch people for their lives to see what happens. The Ethical landscapes and safety landscapes, and regulatory landscapes are all evolving. We have to do rigorous pathogen screening and get consent, and think about animal welfare.

Xenotransplantation heralds abundance over scarcity. Gene-edited pigs could supply organs on demand, equalizing access, reducing black-market risks, and saving lives globally. Early trials in 2025–2026 mark the tipping point—from experimental to potentially routine—offering a future where no one dies waiting.

AlphaFold and AI-Driven Protein Structure Prediction

The screen turns on in a lab and shows a colorful ribbon moving around. A protein takes its perfect shape. This is something that used to take years to figure out with tools like crystallography or cryo-electron microscopy. Now AlphaFold can do it in a few seconds. AlphaFold is a computer system made by Google DeepMind. It has solved a problem in biology: finding out what shape a protein takes based on the tiny building blocks it is made of, which are called amino acids. The protein's shape is very important. Alphafold can find it out from the amino acid sequence alone.

Proteins are like the machines that make life work. They do things like help with reactions and fight off bad things that get into our bodies. Some proteins even help our cells talk to each other. The thing about proteins is that they have to be in a specific shape to work properly. If a protein gets misshapen, it can cause problems, like Alzheimer's disease or cystic fibrosis.

For a time, it was really hard and expensive to figure out what proteins look like. So we only knew what a small number of proteins looked like. Then something new came along called AlphaFold.

AlphaFold 2 was released in 2020. This thing was a deal. It did better than everyone in the CASP14 competition. AlphaFold 2 was really good at figuring out what proteins look like. It used something called learning. This means it looked at information from lots of proteins. Then it used this information to guess how the atoms in a protein are arranged. It kept guessing and checking until it got really close to the answer. The result was that AlphaFold 2 could figure out where all the atoms in a protein are, which is much of what you would get if you did it in a lab. AlphaFold 2 is really good at predicting the shape of proteins.

By 2024-2025, AlphaFold 3 changed things a lot. This happened when DeepMind and Isomorphic Labs launched AlphaFold 3 in May 2024. Now AlphaFold 3 can do more than just predict proteins. AlphaFold 3 can also predict things like complexes with DNA, RNA and small molecules like drugs. It can even predict ions and modified residues. This means AlphaFold 3 covers most of the Protein Data Bank.

The accuracy of AlphaFold 3 is really good. It is at least 50% better at predicting how proteins and small molecules like drugs interact. In some cases, it is 200% better. AlphaFold 3 can show how drugs bind to targets. It can reveal pockets, allosteric sites and how strong the interactions are.

The wonder of AlphaFold is really deep. AlphaFold is helping us understand molecules at a scale we have never seen before. The AlphaFold Protein Structure Database now has all known proteins in it. Over 200 million structures that anyone can look at for free. And this is helping people discover new things in biology. Researchers are now sending in 50% experimental structures to databases, and they are using AlphaFold predictions as ideas to test or as a starting point. By the end of 2025, more than 35,000 research papers will be talking about AlphaFold, and these papers will be about everything from the study of viruses to synthetic biology. AlphaFold is really making an impact on our understanding of biology, and it is helping us learn more about proteins and how they work.

Applications are really changing the way we design drugs. The old way of doing things takes a long time, like 10 to 15 years and a lot of money, billions of dollars. Alphafold is making the early stages much faster. It can find the parts of a protein that we can use to make drugs in proteins that we thought were impossible to work with, like KRAS, which is a protein that can cause cancer or proteins that are really hard to understand.

Isomorphic Labs, a company that came from DeepMind, is using AlphaFold 3 to discover drugs using artificial intelligence, and they are working with big pharmaceutical companies. Some new companies, like Recursion with their Boltz-2 model and Genesis Molecular AI(Pearl), are also using AlphaFold to predict how well drugs will work with proteins. They can do it in a way that is interactive. AlphaFold is really helpful with antibiotics. It helps us design things to fight against bacteria that are resistant to antibiotics. This is done by making models of enzyme inhibitors. AlphaFold also helps with cancer. It makes targeted therapies work faster by showing us how oncoproteins interact with each other. This is very useful for people who are working on cancer treatments. AlphaFold is a tool for making discoveries in the field of antibiotics and cancer.

Custom enzymes are being made for industry. These bio-engineered catalysts can be used for biofuels, breaking down plastic, or green chemistry. We are getting an understanding of diseases. AlphaMissense is a tool that looks at millions of variants and figures out if they are bad for us. AlphaGenome is another tool that looks at how cells are controlled. We are also getting closer to being able to simulate cells. The goal is to be able to model what happens in the nucleus of a cell. Custom enzymes and custom enzymes are really important for this kind of work.

By 2026, AlphaFold evolves further—faster, more reasoning-capable models collaborate like research partners. Critics note limitations (e.g., unusual nucleic acid structures), but iterations address them. AlphaFold doesn't replace experiments; it supercharges them, democratizing structural biology and igniting an explosion in molecular innovation.

Base Editing, Prime Editing, and Epigenome Editing: Precision Without the Scars

Think of DNA as a book that has all the information to make and keep your body working. Sometimes, if there is a mistake in this book, like one wrong letter, it can cause big problems like sickle cell anemia or cystic fibrosis. The old tools that scientists use to fix genes, like the CRISPR-Cas9, work like scissors: they cut the DNA in two spots to try to fix the mistake. The cell fixes the cut. This fix can be a little sloppy. This means the cell can make mistakes. The cell can make changes that it does not need to make. The cell can even make bigger problems, like extra cuts in other parts of the genome. The cell can have problems with the genome because of these cuts. The genome is like a blueprint for the cell. Extra cuts can cause a lot of trouble for the cell.

The new tools in this section are really gentle. They fix mistakes without making those cuts in the DNA that scientists call double-strand breaks or DSBs. This means they do not make a cut in the DNA. The new tools, like base editing and prime editing, and epigenome editing, let scientists make changes. They can change one letter. Rewrite a small part or just turn a gene on or off. They can do all this without changing the DNA letters.

1. Base Editing: Changing One Letter at a Time

Base editing is really cool because it is like we are using a pencil to change one letter in our DNA to a different one, and we can do this directly and, in a very clean way, with base editing.

Scientists use a modified CRISPR system that finds the exact spot in the DNA (guided by RNA). Instead of cutting, it carries a special enzyme that changes one base chemically. There are two main types:

1. One changes C to T (or G to A on the other strand).
2. The other changes A to G (or T to C on the other strand).

Together, these cover four common "point mutations" (single-letter swaps) that cause many diseases.

Fantastic Applications:

- **Sickle cell disease:** The thing about sickle cell disease is that it is caused by a mistake in the hemoglobin gene. This mistake is like a letter in a word. Base editing can fix this letter in the hemoglobin gene and make it normal again. When the hemoglobin gene is normal, the blood cells in people with can work the way they are supposed to. Trails are advancing quickly.
- **Cystic fibrosis:** This thing can fix little mistakes in the Cystic fibrosis gene, which is called CFTR. It might even make lungs work better.

- **Other genetic diseases** with point mutations (like some forms of blindness or liver disorders).

2. Prime Editing: Search and Replace – More Flexible Fixes

Prime editing is really something. People usually call it a search and replace tool. This is, like a word processor, but instead of words, it works with DNA.

This thing uses a special system that is like CRISPR to find the right spot on the DNA. It only cuts one part of the DNA, not both parts. Then it uses a helper called reverse transcriptase that acts like a copy machine. A special guide RNA tells it where to look and what to change. The tool makes a copy of the change and puts it onto the DNA, then the cell fixes it. This lets it:

1. Do all 12 possible single-letter swaps (not just four like base editing).
2. Add or remove small chunks of DNA (insertions or deletions).
3. Combine changes in one go.

This thing is really versatile. It does not make big cuts. It can fix a lot of mutations with high accuracy, and it has fewer unwanted side effects. The mutations can be fixed by this thing with high accuracy, and the thing is really good at avoiding unwanted side effects.

Fantastic Applications:

- Fixing complex genetic diseases where the mutation is

more than a single swap (e.g., some forms of Tay-Sachs, Batten disease, or muscular dystrophy).

- Correcting multiple errors in one treatment.
- Early human trials (like for chronic granulomatous disease) show safe, effective corrections in patients.

3. Epigenome Editing: Turning Genes On or Off Without Changing the Letters

The epigenome editing process does not change the DNA sequence. It just. Removes little chemical tags, like sticky notes, that tell the gene what to do. These chemical tags decide if the gene is used or not. The epigenome editing is about adding or removing these tags to control the gene, not about changing the DNA sequence of the gene itself.

Scientists use a special tool called CRISPR that finds a specific spot without cutting anything. They then add some enzymes to this CRISPR system. These enzymes do an important job. They add or remove special marks on the genes. These marks are like methyl groups. When methyl groups are added to genes, they usually turn the genes off. When they are removed, the genes turn on. The good thing about these changes is that they can stay when the cell divides. Also, these changes are often reversible, which means they can be changed back. The best part is that they do not change the actual DNA code of the CRISPR system or the genes. The CRISPR system and the genes remain the same.

Fantastic Applications:

- **Cancer**: Silence overactive cancer genes or turn on protective ones without risking new mutations.

- **Neurological disorders:** Turn down harmful genes in Huntington's or Alzheimer's models, or boost helpful ones in conditions like depression or epilepsy.
- **Therapies**: Sometimes people have chronic conditions that are not going away. With therapies, doctors can turn the gene activity up or down as needed over time. This is very useful when a full and permanent fix for the condition is not available or possible at this time. Reversible therapies can be very helpful in these situations.

Why These Tools Are Really Important To Us?

These three techniques represent the next big leap in gene editing: safer, more precise, and more versatile than cutting DNA in half. They fix the root cause of many genetic diseases gently, with lower risks of side effects. Companies like Beam Therapeutics (base editing), Prime Medicine (prime editing), and others are running clinical trials right now, with some patients already seeing real improvements.

In the coming years, expect even better versions: easier delivery into the body, higher efficiency, and combinations with other biotech tools. Together, they bring us closer to curing hundreds of "incurable" genetic conditions—turning science fiction into everyday medicine, one careful, scar-free edit at a time.

The Cutting Edge: Synthetic Biology, Retrotransposon/Bridge Editing, Fanzor, and the Future Ahead

This last part is about the cool tools in biotechnology. These tools are not just about solving problems; they let scientists design life to change a lot of DNA and make things that are smaller and safer, and more powerful. We will explain each tool in words, then show how the biotechnology tools work together to make great progress. At the end, we will talk about what the biotechnology tools could mean for your future.

1. Synthetic Biology: Designing Life from Scratch

Scientists think of biology like a computer program. They take living things, like cells or bacteria. Change them so they can do new things that they cannot do naturally. Scientists do this by redesigning living things, like organisms, to do jobs that nature did not create. Synthetic biology is really about making living things work in ways. But how it works:

- They write new "code" for cells using DNA letters (A, T, C, G).
- They build artificial genes, pathways, or even whole new proteins.
- These redesigned cells are really useful because they can make medicines for people, clean up bad pollution, for the environment, produce food that we need to eat or create materials that we use every day.

It turns biology into something we can program, just like writing an app. We are no longer limited to what evolution created over billions of years.

Here are some fantastic examples in 2026:

- Bacteria that eat plastic waste and turn it into useful chemicals.
- Engineered microbes that capture CO_2 from factories and make biofuels or plastics.
- "Living factories" that grow spider silk (stronger than steel) or self-healing materials.
- New proteins that are created with the help of intelligence can do some really cool things, like fight bacteria that are resistant to antibiotics or help clean up oil spills.

2. Retrotransposon-Based Editing & Bridge Recombinase Editing: Rewriting Huge Chunks of DNA

These are tools that let scientists make big changes to DNA. They can move parts of the DNA. Add new parts or remove parts that are already there, or even flip big sections of the DNA around. This is a big deal because scientists can do this with thousands or even millions of letters in the DNA without messing it up in a way that is bad for the DNA. The DNA stays safe. Does not get broken in a bad way.

How they work?

- Retrotransposon editing uses nature's own "jumping

genes" (mobile DNA pieces) combined with precise guides. It can insert big pieces of new DNA cleanly and safely.

- Bridge recombinase editing uses proteins and a bridge RNA guide to make big changes to the genome. It can cut out parts, paste in new parts, flip parts around or delete huge sections of the genome. This is like editing a book. Of changing one word at a time, bridge recombinase editing is like editing big paragraphs all at once. The bridge RNA guide helps the special proteins find the spot in the genome to make these big changes, in the bridge recombinase editing process.

Why they're wonderful?

Normal gene editing can only fix one mistake at a time. The gene editing tools can change entire parts of the genetic book all at once. This is really good for diseases like cancer or for building brand new biological systems from the ground up. The gene editing tools are perfect for this because they can change a lot of things in the book at the same time.

What are Fantastic Applications?

- Deleting faulty gene clusters that cause diseases.
- Inserting whole new gene circuits to make cells produce powerful medicines inside the body.
- We are talking about changing parts of the DNA to make new organisms that are not found in nature. For example, we can make plants that grow faster or animals that do not get sick easily. This is like creating custom organisms with abilities, like plants or animals that can do things that normal ones cannot do.

3. Fanzor: The Compact, New Gene Editor

Fanzor is a family of tools that can edit genes. These tools were found in fungi, animals and algae. The thing about Fanzor tools is that they are really small. They are even smaller than CRISPR-Cas9 tools. This is a thing because it means Fanzor tools are easier to get into human cells. Fanzor tools are a lot like gene editing tools, but their small size makes them special.

How it works?

Like CRISPR, Fanzor uses RNA to find the exact DNA spot it wants to change. But it's compact (smaller size = fits better in delivery vehicles like viruses used in gene therapy). Recent improvements in 2025–2026 made Fanzor much stronger and more accurate.

Why it's wonderful?

When we talk about size, it is really good for delivery into the body. The smaller size means it is safer to put inside the body. It cause fewer problems with the immune system, and due to its smaller size, it will work better for treating diseases that are inside living people.

Fantastic Applications:

- Fixing muscular dystrophy by restoring the missing protein in muscle cells.
- Potential treatments for eye diseases, liver disorders, or brain conditions.
- When you use this with tools, it helps make editing really

precise and gentle.

4. How These Tools Work Together for Big Progress

The real magic happens when these tools work together:

- AI (like advanced AlphaFold) designs brand-new proteins that never existed before.
- Synthetic biology builds cells or organisms that use those new proteins.
- Bridge/retrotransposon editing inserts the new designs into big parts of the genome safely.
- Fanzor fine-tunes or adds final precise changes.

This combination is like having a super-intelligent design team: AI thinks up ideas → synthetic biology builds them → large-scale editing installs them → compact editors polish them. The speed and power multiply, creating progress that was impossible just a few years ago.

What This Means for Your Future?

These technologies are getting better and better fast. You might see some amazing things happen in your life. These technologies are changing so much that you will probably see things that nobody thought were possible. These technologies are going to keep getting better. That is going to change a lot of things.

- **Longer, healthier lives:** Treatments that slow aging, repair damaged tissues, or remove senescent (old, harmful) cells → many people could live to 100+ in good health.

- **Personalized health:** Your DNA is sequenced → AI designs custom therapies → synthetic cells deliver medicine exactly when and where your body needs it.
- **Sustainable world:** Engineered microbes clean pollution, grow food with less water/land, create eco-friendly materials (like plastic made by bacteria instead of oil).
- **New cures:** Diseases once thought impossible to treat (complex genetic conditions, many cancers, neurodegenerative diseases) become manageable or curable.
- **Lab-grown everything:** We can grow organs and meat and fabrics and fuels in a lab. This way, we are not hurting animals or the earth. Lab-grown things are a way to be kind to the planet and to animals. We can have lab-grown organs and lab-grown meat and lab-grown fabrics, and lab-grown fuels.

We are gaining the power to redesign life itself. This brings huge hope—but also big questions:

- *Who gets access to these powerful treatments?*
- *How do we prevent misuse (e.g., creating dangerous organisms)?*
- *How do we protect nature and future generations?*

Fair rules, science that's open to all, and people working together around the world will help make sure these great things are good for everyone. The open science and global cooperation will really make a difference when it comes to these wonders. These wonders should benefit everyone. That is why we need responsible rules and global cooperation to make that happen.

Biotechnology today is similar to the days of computers in the 1970s and 1980s. At that time, not many people thought about things like smartphones or the internet or even AI chatbots. Biotechnology is at that point. The tools we are talking about in this chapter are the start of biotechnology. Over the next 10 to 30 years, we might see a world where people do not get genetic diseases very often, and people do not get old as fast, and humans live in a better way with the planet. This will happen because we will learn how to use biotechnology in a kind way to program life. Biotechnology will help us do this.

The future of biology isn't something that happens to us. It's something we are writing together.

7

Quantum Engine of Life : Stealing Eternity

"The cell is not a factory of gears and levers; it is a shimmering theater of subatomic ghosts. We are just now learning how to hear the music." — *Jim Al-Khalili, Life on the Edge*

Before we peer into the "Ghost in the Cell," we must redefine our understanding of reality. In the world of the very small, the rules of common sense dissolve. To navigate this chapter, you must master these five "Spooky" concepts:

1. Quantum Coherence (The "Shared Rhythm")

- **The Physics:** When particles like electrons or photons act in perfect synchronization, behaving as a single, unified wave rather than individual "bullets."
- **The Biological Context**: This is the "Shield." It allows a plant or a cell to maintain a subatomic "stillness" amid the heat and noise of the body. When coherence is lost

(Decoherence), the system "rusts" and ages.

2. Superposition (The "Everywhere at Once")

- **The Physics:** The ability of a quantum particle to exist in multiple states or locations simultaneously until it is measured or "observed."
- **The Biological Context:** Think of the Quantum Walk. In photosynthesis, energy doesn't pick one path to travel; it takes every possible path at the same time to find the most efficient route instantly.

3. Quantum Tunneling (The "Ghost Walk")

- **The Physics:** A phenomenon where a particle disappears on one side of a solid energy barrier and instantaneously reappears on the other side, without ever actually "crossing" the space between.
- **The Biological Context:** This is how your enzymes work. They allow protons and electrons to "teleport" through chemical walls, making life-sustaining reactions happen trillions of times faster than classical physics allows.

4. Quantum Entanglement (The "Spooky Link")

- **The Physics:** A state where two particles become so deeply linked that they share the same fate. What happens to one happens to the other instantly, regardless of the distance between them.
- **The Biological Context:** This is the "Robin's Compass." Entangled electrons in a bird's eye react in unison to the

Earth's magnetic field, allowing the bird to "see" the invisible geometry of the planet.

5. Wave-Particle Duality (The "Shimmering Identity")

- **The Physics:** The fundamental truth that every subatomic "thing" is both a solid particle and a vibrating wave at the same time.
- **The Biological Context:** This is the "Vibration of Smell." We don't just "touch" molecules with our receptors; we "listen" to their wave-like vibrations. It is the shift from seeing the body as Matter to seeing it as Music.

The Ghost in the Cell

Alright, let's make this text easier to understand for folks with an IQ of 80!

The journey through the first six chapters has been moving forward step-by-step for Prometheus. We've been learning all about the body and how it works. The Last Prometheus showed us that the body is kind of like a car engine. We figured out what all the different parts do and how the whole thing runs. We even learned how to fix problems inside the body.

There's this cool tool called CRISPR that's like a pair of scissors. We can use it to cut out the bad parts and make the body better. The Last Prometheus really helped us understand the body and how to improve it. Now, the cell is kind of like a factory to modern biologists. It's got all these predictable, mechanical parts - gears, levers, and chemical models. BUT when we look REALLY close, past the tiny proteins and atoms, that factory idea starts to fall apart. Those solid gears and levers we thought we had all figured out? They're actually kind of blurry and don't always move the way we expect. Turns out, the cell isn't a factory at all - it's a QUANTUM ENGINE! We're talking about the cell as a whole new kind of thing.

To really understand this "Steal Eternity" thing, we've got to go past the world of biology and engineering and get into the "Middle Kingdom." That's where the crazy stuff happening at the atomic level meets the real world of living things. We're trying to find the "Ghost in the Cell" - it's like a mystery we're trying to solve.

1. The Death of the Ball-and-Stick Model

Alright, let's break this down in a way that's easy to understand. For a long time, biologists have been using these fake models of molecules - they're like little balls connected by sticks to show how the tiny atoms fit together. This makes it seem really simple and straightforward, right? Like if we just know where all the little balls are, we've got the whole living thing figured out.

A lot of people like this idea because it's comforting to think we can understand everything about life if we just know the parts. But the truth is, molecules aren't really like those ball-and-stick models at all. They're way more complicated, and they don't always behave the way we expect.

When we start looking REALLY closely at the tiny stuff that makes up everything, we realize an atom isn't a solid thing - it's more like a cloud where things might be. And an electron isn't just circling around like a little moon, it's actually a wave that can be in lots of places at once!

This "Precision Era" is changing how we think about all this stuff. Atoms and electrons aren't what we thought they were. If life was like a machine, the chemical reactions that keep us alive would be way too slow to actually work. But life doesn't ignore the crazy quantum laws - it actually uses them to its advantage!

It's like life is walking a tightrope, using the weird quantum stuff to get around the slow, clumsy rules of classical physics. We're not just bodies made of flesh and blood - we're actually waves of connected particles from the stars! Pretty wild, right?

The old ball-and-stick models don't cut it anymore. Life is way more complex and mysterious than that. We've got to keep an open mind and really dig into the quantum world to understand how life REALLY works.

2. Schrödinger's Prophecy: What is Life?

Back in 1944, when the whole world was at war, there was this physicist named Erwin Schrödinger. He's the guy famous for that cat in a box that's both alive and dead at the same time - pretty wild stuff!

Anyway, Schrödinger wrote this little book called "What is Life?" and it's probably one of the most important works in the history of modern biology. But get this - he wasn't even a biologist! He was a physicist. Schrödinger pointed out this problem with the Second Law of Thermodynamics. It says that everything in the universe is always getting more and more disorganized over time. This "entropy" thing means things just naturally fall apart and get messy, like a house crumbling or a star running out of energy. BUT LIFE is different! Life takes all that chaos and disorder and makes it organized. Life creates order out of disorder. In fact, Schrödinger said life actually FEEDS on entropy to survive. Isn't that crazy? Life and the universe are like complete opposites - the universe gets more disorganized, but life gets more organized.

Schrödinger thought living things have to follow different rules than machines. A steam engine works in a predictable way, but life doesn't work like that at all. He said that DNA, the stuff inside our cells that gets passed down, has to be special - he called it a "crystal." And get this - Schrödinger thought life has to work at the tiny, atomic level to keep going. That's where these special quantum laws come into play. They help life keep its shape and stay alive, even when everything around it is trying to fall apart. So Schrödinger's big idea is that life uses these quantum rules to fight against all the noise and disorder of the universe. Pretty mind-bending stuff, right? Life is way more complex and mysterious than we ever imagined!

> *"Life seems to be electrochemical at its base, but quantum-coherent at its core." — Erwin Schrödinger*

This is where the idea of the Quantum Architect started. Erwin Schrödinger figured out that if we want to live, which he called Steal Eternity, we need to know how life keeps its Quantum Order in a world that is always trying to mess it up. The Quantum Architect mindset is really important here because it helps us understand how life does this. Quantum Order is what keeps everything together.

The impact of Schrödinger's "What is Life?" book cannot be overstated. It was like a siren song that drew in some of the greatest physicists of the 20th century, getting them to shift their focus from studying dead atoms to studying living matter.

Take James Watson and Francis Crick, the two scientists who figured out the structure of DNA. They said they decided to stop studying physics and start studying biology because of Schrödinger's book. They were really intrigued by this "aperiodic crystal" that Schrödinger had described as the key to life.

You know, most people think Watson and Crick kicked off the Information Age of biology when they cracked the DNA code. But Schrödinger was aiming for something even bigger. He didn't just want to find the code - he wanted to uncover the quantum mechanism that allows that code to stay the same for thousands of years!

Even 80 years later, we're just starting to really dig into what Schrödinger was talking about. The Information Age he inspired is important, no doubt. But Schrödinger's deeper ideas about the quantum underpinnings of life are still crucial. We've barely scratched the surface of what he envisioned.

It's amazing how one physicist's "small book" could have such a profound and lasting impact on the course of modern biology. Schrödinger really was a visionary, wasn't he? His ideas continue to shape the way we think about life at the most fundamental levels. We've got a lot more to uncover, but Schrödinger laid the groundwork for it all.

4. The "Warm, Wet, and Noisy" Paradox

For decades, mainstream science dismissed the idea of "Quantum Biology." Physicists argued that quantum effects like Superposition (being in two states at once) and Entanglement (instantaneous connection between particles) were incredibly fragile. To be able to see these things happening in a laboratory, you have to make the atoms very cold, almost as cold as they can possibly get, which is -273.15 degrees Celsius, and you have to put them in an empty space, which is called a vacuum. Quantum Biology is an interesting idea, and scientists are still trying to learn more about Quantum Biology and how it works.

The inside of a cell is really different. It is 37°C, which is the same as 98.6°F. The human cell is a place, and it is very crowded. There are a lot of molecules that are always vibrating. That makes it very "noisy".

When we think about the cell in terms of standard physics, the quantum coherence (a fundamental quantum mechanical property where particles (like electrons or photons) maintain a fixed, predictable phase relationship, allowing them to exist in multiple states simultaneously (superposition).) of the human cell should disappear very quickly. This happens in a fraction of a billionth of a second. This process is called Decoherence of the cell.

Life is the thing in the universe that knows how to keep the Quantum Light on. And it does this in a room that's really not perfect. It is Warm, Wet and Noisy. I mean the Quantum Light stays on in this Warm, Wet and Noisy room because of Life.

Life is really good at using proteins to protect its quantum states from the things around it. It takes the heat. Uses it to create a kind of rhythm that actually helps the tiny particles in quantum move better. Life is like a well-made machine that connects the really small things to the big things we can see. Life does this by using these proteins to make the heat work, for it is not against it, and that is what makes life an engineering masterpiece that brings together the tiny world of atoms and the big world of things we can see and touch.

5. Interesting Fact: Your Eyes are Quantum Detectors

You don't need a multi-billion dollar laboratory to see quantum biology in action. You just need to walk into a dark room.

Researchers have proven that the human eye is sensitive enough to detect a single photon—the smallest possible unit of light. A photon is a quantum object. When that single photon hits a protein in your retina called Rhodopsin, it triggers a chemical change that sends a signal to your brain.

This means that your biological hardware is calibrated to the absolute "floor" of the universe. You are equipped with sensors that can detect the transition from the quantum world to the classical world. You aren't just "seeing" light; you are interacting with the fundamental "packets" of reality. If your eye can do it, what other quantum secrets is your body hiding?

6. The Scale of the Strange: Superposition and Tunneling

To understand the sections ahead, we must define the two "Ghostly" powers that life utilizes:

Quantum Tunneling: Imagine a ball trying to get over a hill. In the classical world, if the ball doesn't have enough energy, it rolls back down. In the quantum world, the ball can simply "vanish" on one side of the hill and "reappear" on the other. It walks through the wall. As we will see next, your enzymes use this "teleportation" to make life-sustaining reactions happen at the speed of thought.

Coherence (Superposition): This is the ability of a particle to test every possible path at once. It's like a GPS that doesn't just calculate a route, but actually drives every possible road simultaneously to find the one with the least traffic. Plants use this to harvest sunlight with almost 100% efficiency.

7. Conclusion: The Middle Kingdom

The "Ghost in the Cell" is the bridge. It is the realization that we are not solely governed by the laws of "Big Things" (Newton) or the laws of "Small Things" (Bohr). We exist in the Middle Kingdom, where life has spent four billion years learning how to harvest the spookiness of the atom to overcome the gravity of death.

If we want to steal eternity, we cannot simply be geneticists; we must become Quantum Tuners. We must learn how to maintain the "Coherence" of our cells against the "Decoherence" of aging. We must learn how to keep the "Ghost" in the machine.

The Robin's Compass: Entanglement in Flight

If the Ghost in the Cell is like the software that makes life work, then Quantum Magnetoreception is the advanced hardware that life uses to interact with the world.

For a time, people who study nature were amazed by the European Robin. The European Robin can fly from the forests of Scandinavia to the warm places in Africa. This is a long way. People thought the European Robin was just following its instinct. When scientists say instinct, they really mean they do not understand how something works. The European Robin and Quantum Magnetoreception are still not fully understood. Quantum Magnetoreception is a part of the Ghost in the Cell that helps the European Robin navigate.

When the we turned the microscope into a quantum probe, we found out that the robin is not just following a gut feeling.

The robin is actually using what Albert Einstein called "Spooky Action at a Distance". The Last Prometheus showed us that inside the eye of a migrating bird, like the robin, there is a compass. This compass is very sensitive. It can detect the Earth's magnetic field. The Earth's magnetic field is a weak force. It is 100 times weaker than the magnet that you have on your refrigerator.

1. The Cryptochrome: A Quantum Lens

The secret to this is a protein called Cryptochrome that is found in the retina of the bird. When one photon of blue light hits Cryptochrome, it moves an electron from one molecule to another molecule. This makes something that physicists call a Radical Pair. The bird has Cryptochrome in its retina.

These two electrons are not just close to each other; they are Entangled. This means what happens to one electron will happen to the electron at a really small level. If you do something to one electron, the other electron will be affected too. Because these electrons are entangled, they exist in a fragile state that can be easily changed by things outside of them. Specifically, Entangled electrons are sensitive to the Angle of the Earth's Magnetic Field.

When the bird turns its head, the magnetic field makes a change to the spin of the electrons that are connected. This tiny change affects the way the bird's eye sends signals, and scientists think the bird sees this as a kind of shadow or pattern, on top of what it sees. The bird is not thinking about which way is north; it is actually seeing the shape of the magnetic field that surrounds the earth, the magnetic geometry of the planet, and that is what guides the bird.

The amazing evidence for this quantum compass thing did not come from seeing birds fly around outside, but from putting them in a special kind of cage that blocks radio signals. Back in 2004, a researcher named Thorsten Ritz did an important experiment. He took some migrating robins. Put them in little wooden houses where they could easily figure out which way to go to get to where they were migrating to. The migrating robins could see their path. Then he put these houses inside a special thing called a Faraday Cage. This Faraday Cage is like a mesh that stops all the noise from getting in. You know what happened. The migrating robins were still able to navigate fine. The migrating robins navigated perfectly.

The fact is that when Ritz experimented with a weak radio signal, the birds got totally confused. This signal was so weak that it would not even mess up a phone call. The birds completely lost their sense of direction. They started hopping around in circles. Had no idea which way was North. The birds were basically blind to the direction they needed to go. Ritz found out that the birds were really affected by this radio signal.

The idea here is that this discovery confirmed that the compass is related to things. Normal magnets, like the ones you find in a compass, are not affected by small amounts of radio frequency noise but Quantum Entanglement is affected by it. The radio frequency noise messed up the connection between the electrons in the birds' eyes. We figured out that our modern world, which is full of all the noise from our devices, might actually be stopping the old quantum sensors of the animal kingdom from working properly. The Quantum Entanglement in the birds' eyes is really sensitive to this kind of noise.u

A robin's compass is really weird because it only works when the bird can see the Blue Light. The thing that makes it work is that the robin has to be able to see this Blue Light. It is like the Blue Light is what makes the robin's compass do its job. The Robins Compass will not work at all if it cannot see the Blue Light. The thing about robins is that if you put a migrating robin under a red light, it gets completely lost. This happens because the robin needs a kind of light to navigate, and that light has to be from the blue part of the spectrum. The blue light is what helps the robin figure out where it is going. Without light, the robin's special navigation system does not work. The robin can still see things with its eyes. It cannot find its way because its internal navigation system is not working. The migrating robin has a way of knowing where it is going, but the red light messes that up. This is a reminder that biological magic is really a very specific kind of engineering that uses light to work with tiny things that are too small to see, like the things that quantum engineering is about. Biological magic is actually activated quantum engineering; it is not just something that happens by itself, it is something that is made to happen by the way that light and tiny things work together in biological magic.

The idea behind Robin's Compass is that Life has been doing Quantum Sensing for a long time. Millions of years. We think we are the people to work with Quantum things because we build big computers in expensive labs that cost a lot of money. The robin shows us that nature was doing this a very long time before we were. To a robin, the Earth's magnetic field is not something that affects the physical world; it is also Information that the robin can use. The robin uses the Earth's field to find its way around, which is a kind of Quantum Sensing that the robin does naturally. Life and the robin have been doing Quantum Sensing for a long time. We are just now starting to understand how it works with our Quantum Architects and big computers.

The robin does not have a soul that knows the way to Africa. It has something that reads the information of the planet. This shows us a beautiful truth. The universe is always sending out its location, and life is a bunch of things that have developed the right tools to hear it. The robin is one of these things. If we want to live, we have to stop thinking of the body as just a collection of parts, like the heart and lungs. We have to start thinking of the body as a thing that gets the signals of the universe. The universe is always sending out these signals, and the robin is one of the things that can hear them. We need to think of our body as a receiver, like a radio, that can get these signals from the universe.

As we get closer to 2027, we are learning a lot from the Robin's eye. The things we learn from the Robin's eye are already being used to make technology for people. Engineers are trying to make sensors that work like the cryptochrome in the robin's eye. These Bio-Inspired Quantum Sensors will help us make navigation systems that do not need GPS. This means these navigation systems will keep working even if someone tries to stop them or send signals to them. The new navigation systems will be, like the Robins navigation system, which always knows where it is going.

For the Architect, this is a big deal. If a tiny thing like a protein in a bird's eye can stay connected, in a way, even when it is warm and wet, then that tells us something. The thing that makes our bodies get old and fall apart, which is called Decoherence, is not something that always has to happen. The Architect sees this as a problem that can be solved, a challenge to figure out. The Robin is a bird. If the Robin can stay in a state called a quantum state and keep it coherent, that means it can find its way to Africa. Maybe we can learn from Robin. Maybe we can figure out how to keep our cells in a coherent state. If we can do that, maybe we can find a way to live past the point where our bodies normally stop working, which is what we call death. The Robin finding Africa is really about the Robin. Our cells finding a way past death is really about us and our own bodies.

The Chlorophyll Computer: Perfect Coherence

European Robins use quantum mechanics for navigation, while grass uses quantum mechanics to produce energy for the entire world. The "Last Prometheus" story shows photosynthesis as the greatest energetic theft. The sun's chaotic, highly energetic photons undergo transformation through this process into chemical sugar. For decades, we thought this was a simple chemical reaction. The assumption was incorrect.

Standard physics suggests that when light hits a leaf, the energy should bounce around like a pinball, losing heat at every turn, until it luckily stumbles into the "reaction center" to be processed. The photosynthesis process would demonstrate high inefficiency if the assumption proved true. Nature operates through mechanisms that achieve maximum efficiency. A Berkeley team discovered in 2007 that plants achieve energy transfer efficiency close to 100% through their utilization of Quantum Coherence.

The "Quantum Walk" to the Exit

The problem is one of speed versus decay. The FMO Complex light-harvesting protein receives a photon, which triggers the creation of an Exciton that becomes a packet of energy. The exciton has a brief lifespan, since it must reach the "reaction center" within a few picoseconds to convert into fuel or else it will become useless thermal energy.

The classical world requires the exciton to follow a "Random Walk," which involves moving between chlorophyll molecules in the same way a drunk person attempts to navigate through a dark forest. The person in the dark forest would become lost and die before discovering the exit.

The plant executes a Quantum Walk instead of executing a random walk. The exciton exists in Superposition because it functions as a wave-function. The object travels multiple paths at the same time instead of choosing between two options. The system "samples" all routes while the Constructive Interference process intensifies the paths that reach the exit point while eliminating the paths that stop at dead ends. The plant "finds" the exit because it is everywhere at once.

The "Green" Quantum Computer The discovery of quantum coherence in plants was so shocking that many physicists initially refused to believe it. Graham Fleming and his team "watched" energy movement through green sulfur bacterium by using ultra-fast laser pulses (femtosecond spectroscopy) in 2007. The researchers observed synchronizing "Quantum Beats" that developed as indications of coherence. The first instance of "Spooky Action" was documented when scientists recorded it in a living organism at room temperature.

Every tree leaf serves as a Biological Quantum Computer. The system operates through probability calculations, which leads to the determination of optimal energy transport pathways. Your backyard ferns have operated silicon-based quantum computers for three billion years, while we have tried to develop them for thirty years.

The "Noise" is the Music. The "noise" that humans encounter during quantum computing represents one of the main difficulties because the nearby atomic vibrations cause the quantum state to "collapse." We attempt to stop vibrations by freezing our computers. Plants do the opposite. Research has shown that the protein "scaffolding" holding the chlorophyll molecules vibrates at a specific frequency that actually maintains the quantum coherence. The plant uses vibration to keep excitons "shimmering" in superposition for longer, instead of the vibration "breaking" the quantum state. The "noise" operates as a system feature instead of being classified as a system defect. The leaf functions as a "tuner," which aligns with the sun's subatomic frequency.

We came to realize that Nature does not gamble; it calculates. The world existed as a realm of "Trial and Error" until this point in our history. Life has advanced from "trying" to "knowing" because photosynthesis exists at the fundamental level.

The plant achieves "friction" elimination through its use of quantum walk possibilities, which extend beyond physical boundaries. Life achieved survival on this planet by discovering how to use the atom's "Ghost" to violate the Second Law of Thermodynamics. We must abandon all energy-exertion attempts in order to attain "Steal Eternity"; we need to master "energy flow" through all accessible paths.

The "Last Prometheus" project aims to replicate the plant's

homework from its original creation towards the 2030s. Scientists are designing Quantum-Coherent Solar Cells that mimic the FMO complex. If we can master this "Subatomic GPS" for energy, we could create power grids with zero loss and batteries that never get hot. The Architect's ultimate objective involves achieving Internal Energy Mastery. Increased body energy efficiency would occur through mitochondrial protein "tuning," which maintains body coherence. We are looking for the "Green Key" to an eternal engine.

The Tunneling Catalyst: The Magic of Enzymes

The enzyme functions as a "lock and key" mechanism in conventional biological processes. The process involves a specialized protein that attaches to a molecule. This protein functions by holding the substrate in a specific orientation to reduce the "activation energy" needed for chemical reactions. The coffee you consumed today would need years for digestion without enzymes, while your lungs would not be able to deliver oxygen to your bloodstream. The "lock and key" mechanism encounters a mathematical problem because the calculations do not produce correct results.

Some biological reactions happen trillions of times faster than they should, even with a catalyst. The molecules show behavior that resembles their movement via bridge crossing because they skip the bridge and reach the opposite bank. The "secret engine" of living cells operates through Quantum Tunneling (a phenomenon in which a particle passes through an energy barrier it classically lacks the energy to overcome), which governs its essential functions. The body of the "Last Prometheus" requires tunneling control for its preservation because it enables the body to remain functional without experiencing the effects of bodily decay.

In the classical world, if you want to get over a hill, you must have enough energy to climb to the top. The boundary confines your movement to your own space. The quantum realm shows that particles like electrons and protons display dual characteristics as both particles and waves. A wave doesn't have a single, solid location; it has a "tail" of probability that extends through barriers.

If a barrier is thin enough—like the gap between two molecules inside an enzyme—the "tail" of a proton's probability can reach all the way to the other side. There exists a probability that the particle will disappear from its original location to reappear directly at the target position. The object has not "traveled" through the space because it has "tunneled" through the energetic barrier.

The most famous evidence for quantum tunneling in biology comes from the study of Enzyme Kinetics at different temperatures. In the 1980s, researchers led by Judith Klinman at UC Berkeley began looking at enzymes that move hydrogen atoms (protons). The classical chemistry theory states that colder temperatures should cause reactions to slow down because there is not enough heat energy to push the atoms over the hill.

The scientists discovered that specific enzymes needed for life showed a "plateau" point of reaction performance. The performance of the system reached its lowest value, which remained unchanged as the temperature decreased towards freezing. This became the "Smoking Gun" evidence. The particles were tunneling because they continued to react without any heat energy available to move them. Life had developed a "Quantum Shortcut" that functioned in the icy, silent world of subatomic particles.

How do we know it's tunneling and not just really good "mechanical" pushing?Scientists can replace the hydrogen in a lab-grown organism with Deuterium ("Heavy Hydrogen"). Deuterium atoms possess double the mass of standard hydrogen atoms, yet maintain identical chemical "shape." The classical "lock and key" system indicates that the reaction rate would decrease as the atom's weight increased.

In the quantum world, however, the "wavelength" of a heavier particle is much shorter, making it much harder for it to "tunnel" through a wall. "Heavy Water" research shows that the enzymatic reactions stop working because their speed falls below 100 times, or even more, of their original speed. The specific enzyme design proved this because it required light hydrogen's "wave-like" properties to function. Your body operates as a precisely balanced tool that has been "weight-calibrated" to transport protons through teleportation.

The process of "Stealing Eternity" requires us to abandon the belief that we can use drugs or surgeries to force our bodies into good health. The Quantum Landscape of our proteins requires us to master its engineering process. We can accelerate life processes while improving their efficiency and reducing metabolic "smoke" through our ability to control enzyme barrier "width" modifications at nanometer-scale precision. Our work involves enhancing a teleportation network while we restore functionality to a machine.

Quantum Medicine aims to develop synthetic enzymes that achieve advanced "tunneling" abilities throughout their operations. The "Engineered Enzyme" possesses the capability to tunnel a proton through a viral protein, which will enable it to destroy the protein while being "out of tune" with human proteins.

The Quantum Mutation: When Protons "Slip"

The Chapter 5 section treated DNA as a digital code system that functions like a portable document format which allows vertices to be precisely cut and folded. The Last Prometheus movie shows that life exists through a subatomic engine because life exists between two permanent states. Protons make up this material and protons exhibit a behavior where they tend to leave their assigned locations according to our earlier findings. Scientists need to protect themselves from quantum biology threats because they have proved that evolution occurs through random quantum events. The double helix structure of DNA carries a built-in error mechanism which creates "ghostly" mistakes whenever genetic mutations arise from environmental agents such as UV light or harmful substances. The "spookiness" of an atom leads to a Quantum Mutation which causes a genetic defect that "molecular scissors" cannot eliminate.

The Tautomeric Shift: The Teleporting Letter:

To understand this, we must look at the "Rungs" of the DNA ladder. The four bases A T C and G bind together through Hydrogen Bonds. These bonds represent a "catch" game that centers on the movement of a single proton. The proton usually remains attached to one side of the ladder which guarantees that A will always match with T while C will always match with G.

The Glitch Per-Olov Löwdin proposed in 1963 that a proton would "vanish" from its correct position on one base through Quantum Tunneling but would "reappear" on its partner base during the DNA unzipping process.

The chemical shape of the DNA letter changes for a billionth of a second. The DNA letter becomes unrecognizable at that moment when polymerase passes through. An "A" suddenly looks like a "G." The machine inserts an incorrect letter into the new strand which establishes a permanent mutation. The Ghost has changed the script.

The Physics of the Perfect Mistake:

The scientific community considered Löwdin's theory to be "fringe" for several decades because physicists believed that the nuclear environment produced too much "noise" for protons to successfully execute clean tunneling. The story describes how researchers achieved their goal by using advanced computer modeling to recreate the DNA double helix environment at the University of Surrey.

The protons which were required for base movements between bases needed to overcome "hills" which had lower heights than expected. The findings showed that proton tunneling behaves like a continuous "freak accident" which operates as background noise throughout our daily lives. The researchers discovered that quantum tunneling functions as a main cause for "spontaneous mutations" which result in cancer and aging. The universe generates fundamental uncertainty which makes us victims to both our environment and its inherent nature.

Chemical-induced mutations usually maintain their presence at the original location. Quantum mutations occur during a brief moment which results in their complete disappearance after that time.

The proton exists in Quantum Equilibrium which allows the "typo" to appear and disappear within an instant. The daughter cell receives a permanent mutation from the "flash" DNA copying process which happened during the original cell's DNA "snapback" to normal state. We could sequence a parent cell and find no error yet the "child" cell is born with a life-altering disease. The Genetic Architect faces their most challenging forensic investigation when they deal with a crime that has no evidence other than a spectral echo..

Life encounters these quantum slips according to Jim Al-Khalili who claims that evolution utilizes main paths as its path to development. The evolution process needs quantum typos because they serve as fundamental material for the process. The human race has evolved from subatomic glitches. The choice between fixing quantum noise or accepting Uncertainty will decide our path to "Steal Eternity."

The Architect's Shield Quantum Error Correction

Quantum Error Correction (QEC) for the genome will become the leading research method in longevity studies by 2026. The nucleus environment needs adjustment to reduce tunneling which makes it impossible to stop protons from tunneling.

Scientists investigate Quantum Stabilization methods which involve utilizing electromagnetic frequencies to create a locking effect that secures hydrogen bonds during DNA replication. We set out to create a Subatomic Shield which will defend our identity by keeping our personal Information secure from quantum world forces.

Warm, Wet, and Noisy: The Coherence Shield

In the "Precision Era," the greatest technological hurdle for the "Last Prometheus" isn't building a faster processor—it's keeping a quantum state from "collapsing." To build a quantum computer, companies like IBM and Google must cool their chips to temperatures colder than deep space (0.015 K) and house them in vibration-proof vaults. One stray photon, one tiny bump, and the "spooky" coherence vanishes.

You exist in this moment. Your body exists as a warm, squishy, vibrating mass of 37°C meat throughout the day while you receive constant exposure to light and sound and water movements. Your "Quantum Engine" should have stopped working according to all standard physics laws before your birth. The Coherence Paradox creates a question about how life can maintain "subatomic stillness" in a world that presents loud sounds and high temperatures, and messy conditions.

Biological structures (like the proteins in your brain and the chlorophyll in a leaf) have developed their capacity to synchronize with environmental sounds instead of trying to escape these noises. The "scaffolding" of proteins that surrounds our quantum bits produces vibrations that create a protective barrier through its rhythmic movement because the vibrations match environmental frequencies. The process creates a "shield" which protects against chaotic decoherence to enable quantum states to exist for 1000 times longer periods than they would in a test tube.

The "Quantum Consciousness" Controversy:

The scientific community accepts that our thoughts create the "Warm, Wet, and Noisy" shield, which exists in our enzymes, according to the most extreme scientific theories of modern science.

In the 1990s, physicist Roger Penrose and anesthesiologist Stuart Hameroff proposed the Orch-OR (Orchestrated Objective Reduction) theory. The scientists showed that microtubules function as critical components for neuron structure, which enables their operation as quantum computing devices. The scientists claimed that consciousness emerges when quantum coherence spreads throughout the brain.

The scientific community mocked the researchers for twenty years because they believed the brain could not operate quantum effects because it contained excessive moisture. The 2014 research conducted by Japanese scientists proved that microtubules can execute high-frequency "vibration-linked" coherence at room temperature. Although we have not shown that your "soul" exists as a quantum wave, we provide evidence that your brain functions as a "Quantum Shielded" environment.

The biological process of aging begins when "Decoherence" (the loss of quantum order) occurs at the subatomic level, and the rate of this process determines your expected lifespan.

The Fact that the Bowhead Whale can live up to 200 years demonstrates that some long-lived organisms develop more stable environments for their cells to function. Scientists are exploring whether these animals developed unique "Heavy Isotopes" or specialized protein "shocks" to create defenses against quantum decoherence. The organisms use subatomic vibration stabilization to achieve a state of bodily operation that decreases the passage of time according to the universal understanding of time. The beings demonstrate extreme physical durability while maintaining a state of silence at the subatomic level.

Nature possesses superior physicist skills compared to our abilities. People attempt to combat the surrounding environment, while life exists in harmony with nature. Humans perceive "Heat" and "Noise" as obstacles to achieving exact measurements, yet life employs them as dual functions that sustain the quantum process.

Life exists between the two realms through its existence at the boundary that separates the "Quantum World" from the "Classical World." The device functions as a Transducer. The system transforms the random vibrations of water and solar energy into a "Coherent" stream of life. The path to "Steal Eternity" requires us to abandon our efforts to construct "Perfect Labs" and begin acquiring the knowledge needed to create "Perfect Filters."

The "Last Prometheus" will expand its capabilities in 2026 through its transition from gene editing to the practice of coherence engineering. We are creating synthetic biological structures that function as "Subatomic Shock Absorbers" through our development of Metamaterials.

The cellular shield of our body will be strengthened through these injections to defend our neurons and heart cells against "quantum leak," which causes cellular damage. We are attempting to make the "Warm, Wet, and Noisy" human body the most stable quantum computer in the galaxy. The "Shield of Prometheus" construction is currently underway.

The Vibration of Smell: Olfaction's Quantum Secret

Biology textbooks have used the "Lock and Key" theory of smell as their only explanation for decades, which states that nose receptors function as locks that can only be opened by specific odor molecules that act as keys. The body transmits a signal to the brain when a molecule of precise shape binds to the receptor; this enables you to identify both "rose" and "rotten egg" scents. The explanation appears to be mechanical and straightforward, yet it contains a significant flaw.

There exist molecules that have identical shapes yet produce entirely different scents, while other molecules with distinct shapes create identical aromas. The "lock" does not explain the "music" which follows. In 1996, biophysicist Luca Turin proposed a radical alternative: we don't "shape-match" smells; we listen to them. Our noses conduct Electron Tunneling Spectroscopy experiments.

Every chemical bond produces vibrations at distinct frequencies that resemble the sound of a guitar string. The Carbon-Hydrogen bond produces a different sound than the Carbon-Sulfur bond. Turin's theory claims that an odor molecule enters your nose and binds to a receptor while an electron attempts to tunnel from one side of the receptor to another. An electron can only make the jump when the molecule located in between produces its exact vibrational frequency needed for the jump. Your nostrils do not examine the molecular structure; instead, they detect its atomic vibration patterns.

The "Heavy" Fruit Fly Experiment

The researchers needed to discover two molecules that had identical shapes yet produced different vibration patterns. They turned to Isotopes. Scientists took a scent molecule and replaced all its normal Hydrogen atoms with Deuterium (Heavy Hydrogen) to create a new version of the molecule. The "Lock and Key" receptor sees the shape as identical to its original form. The Deuterium atom produces a completely distinct sound to "Vibration" receptor systems because it weighs twice as much and its vibrations move at a slower speed. Fruit flies in a well-known study learned to stay away from one particular odor. The fruit flies detected the difference between the two forms of that same odor when they were given the "Heavy" version. The flies could use their hearing ability to differentiate between two objects that had identical shapes but different weights. Quantum smell research reached its decisive evidence point because all living beings sense atomic frequency patterns instead of their external shell structures.

The Symphony of Selective Perception

The intellectual shift here is profound: The nose functions as a biological spectrometer. Turin's theory requires us to stop classifying our senses as "detectors of matter" and to begin seeing them as "energy interpreters." The way we perceive reality functions at a "musical" level, which remains hidden from most people. We experience our environment through a collection of overlapping sound waves instead of solid objects that contact our sensory systems. The "Lock and Key" model is a 19th-century mechanical dream, but the "Vibrational" model is a 21st-century quantum reality. The brain operates as more than a data-processing computer; it functions as an orchestra conductor who oversees a complete subatomic performance. We need to learn how to adjust our biological strings because this skill allows us to maintain our internal sensory balance while avoiding the disruptive effects that lead to various illnesses.

The Smell of Diamonds

Vibration serves as the basis for smell; therefore, scientists could possibly develop a way to "smell" objects that lack any aroma. Some researchers suggest that the reason we can "smell" the approach of rain or the metallic tang of a clean room isn't just because of particles in the air but because of how the atmospheric charge changes the vibrational baseline of our receptors. We comprehend how the air feels at the quantum scale. Our biological hardware can detect environmental subatomic "reverb" at an extremely sensitive level.

The Quantum Architect: Engineering the Subatomic

The Quantum Architect is the final design of the Last Prometheus at the present point of the Precision Era. The previous chapters delivered to us the ability to function as Genetic Librarians for DNA study through the organization of DNA sequences. We now understand that the books consist of paper that produces vibrating sounds. We can change the word's meaning by controlling the vibration without opening the book.

Quantum Architecture functions as the method that uses

Subatomic Resonance Therapy to create its therapy system. The research shows that enzymes and receptors use quantum tunneling and vibration for their normal functions. The disease occurs because our body systems fail to tune their operations. A cancer cell represents a dissonant cell, while aging tissue becomes a muffled one. The process of healing requires us to use a special tool that functions as a subatomic tuning fork instead of chemical hammering techniques.

The first non-invasive method of editing through resonant healing. The medical field begins to adopt experimental therapies that require fewer surgical procedures than CRISPR operator Surgical Scissors during the year 2026. Scientists use low-energy pulsed electromagnetic fields (PEMFs), which operate at biological machine tuning frequencies, to make small adjustments to biological particles back to their primary quantum operating paths.

When a proton begins to enter a quantum mutation, the Quantum Architect functions as a control system that uses external frequency to make the energy barrier unbreachable by the proton. We construct a Light Scaffolding system that supports the Ghost of the Cell structure.

The complete molecular hammer technology represents the first degree of its basic scientific function. The prototypes of this technology now exist to battle against tumors which medical science cannot treat. In late 2023 and throughout 2024, researchers at Rice University demonstrated what they called Molecular Jackhammers. They took specialized dye molecules that can attach to the surface of cancer cells. The molecule starts to vibrate all its parts in a unified manner when it receives exposure to the special near-infrared light frequency.

This vibration creates such intense power that it breaks the cancer cell membrane, which leads to cancer cell death in a few seconds. This substance functions as Mechanotherapy at the molecular level instead of being a drug. Control over a single molecule's vibration gives you the ability to decide when a cell should live or die. The process of direct surgical operation will reach its first stage through a method that uses light as a surgical tool.

The human perception of mass has already experienced an ontological transition, which led to the current information-based existence. The Quantum Architect shows that Materialism in medicine reached its final termination point. The human body transformation from its Mass paradigm needs to occur through its Information-based system, which treats the body as a network of wave-function interactions. The academic field reached its highest point through this intellectual act of stealing fire from the gods. The human body operates as a system of vibrations, which creates sound, and health exists when all body parts maintain their rhythmic flow. The Architect operates as an Acousmatic Engineer who ensures the organism's Subatomic Song maintains its balance with the universe's basic frequencies. The Quantum Signal requires perfect transmission without any loss to achieve its eternal state.

The human body needs to change its material existence for eternal life, which must be achieved through sound-based existence. The Bio-Photon Telephone generates an interesting scientific fact. The Architect uses the tuning fork to find her path. The human body contains a complete system to establish internal connections with its components.

Cells emit incredibly weak pulses of light called Biophotons. For years, these were dismissed as metabolic noise. Recent studies have found that the detected photons act as a tiny internal laser system. Biophotons serve as the internal language system that cells use to connect at light speed for executing complex tasks such as wound recovery. Light-based communication allows us to establish disease presence approximately three weeks before body infection occurs. The body operates through Quantum Phone Lines, which scientists have begun to explore.

Stealing Eternity: The End of Entropy

The Last Prometheus has finally reached the heart of the labyrinth. We have moved through the genetic code, the biological machine, and the vibrational senses. The Second Law of Thermodynamics stands before us as our final challenge. The law states that entropy must rise indefinitely because it represents disorder. The classical biological perspective explains that aging occurs through the build-up of thermal noise, which causes molecular parts to deteriorate like rust, which cannot be completely fixed. The Quantum Engine allows users to change entropy values instead of following mandatory entropy requirements. Our research showed that through our understanding of subatomic forces, we discovered that living organisms use quantum mechanics to escape the effects of time. The process of "Stealing Eternity" changes a body that operates as a "burning candle" into a body that functions as a "superconducting circuit."

1. Aging as Decoherence:

The Precision Era has introduced new definitions for aging. The process of aging has transformed into Decoherence. Life sustains its life force by protecting its quantum states from external environmental disturbances. The dynamic decoupling system begins to lose its capacity to protect the system after several decades of operation.

Our mitochondria display an energy "Quantum Walk" operation, which reaches maximum efficiency. The enzymes "Tunneling Catalysts" lose their ability to reach target destinations, which causes life processes to slow down. The "Ghost" begins to flicker. The process of aging starts when "Quantum Signal" transitions into "Classical Noise." We need to reverse aging by restoring the original signal instead of extending human life with additional years.

2. The 2025 "Mito-Tune" Trials:

The Mito-Tune clinical trials marked the first successful attempt to use quantum engineering for "Stealing Eternity" in 2025. Researchers designed "quantum-dot" nanoparticles to operate inside the mitochondria of aging human heart cells for their study. The nanoparticles existed to function as Quantum Stabilizers rather than deliver a drug. The system stopped energy diffusion by matching the external FMO complex frequency, which functions as the plant's quantum computer. The trial used stabilizers to treat heart cells from 80-year-old donors. The treatment caused energy production efficiency to increase by 40% within 48 hours, while it returned to levels observed in 20-year-old tissue.

The chemical treatment did not repair our cells because we created Perfect Coherence by returning their subatomic particles to that state. Entropy "Clock" retrograde movement is possible when using a sufficiently advanced screwdriver, according to our demonstration.

The final Promethean Leap occurs when humans master entropy since they gain the ability to experience both Chronos and Kairos. The moment the wave-function fully collapses to a quantum state results in human life passing into the death state. "Eternity" describes an eternal state that prevents temporal decay from occurring. The Architect uses quantum coherence maintenance to take the organism away from experiencing "Thermodynamic Time."

People view death as an inevitable fate, according to ancient philosophers who defined humans as creatures who move toward their death. The Information Wave-fronts have developed the ability to stop their natural dispersal now. The "Quantum Shield" needs to be powerful enough to protect "Soul," which consists of coherent information, to "Steal Eternity." We are operating a Continuous Frequency process, which does not have a conclusion.

The "Immortality" of the Germline:

Nature has already shown us that "Stealing Eternity" is possible.Your somatic cells, which consist of skin and liver, and brain tissues, will age and die, while every human germline cell, which produces children, will retain its immortality. The germs have survived through four billion years of continuous existence, which has maintained their unbroken genetic connection.

Scientists believe that germline cells maintain their youthful characteristics because they possess advanced Quantum Error Correction capabilities. The part that nature considers worthy of preservation as Coherent stands as our only existing coherent element. The Architect aims to extend the germline's "Quantum Immortality" across all body parts. We aim to make "Self" permanent while we make "Species" into eternal elements of existence.

The Architect's Blueprint: The Divine Resonance

The Quantum Engine has reached its final operational point. The chapter started with us examining the "Ghost in the Cell" and ended with our destruction of the Entropy operational system. The "Last Prometheus" no longer considers the human body as an organ system or as a genetic sequence since 2026. We regard it as a Blueprint of Resonance. The Architect's Blueprint represents the ultimate combination of physics and biology. The discovery has been made that when we achieve control over subatomic spaces, we obtain the power to control all visible elements of the universe. Our role has changed from "Gardeners" who prune live branches to "Frequency Engineers" who manage soil vibrations. The laboratory space reaches its divine point through human conduct since we now possess all powers that were previously considered divine. The Unified Theory of Biological Information. The final piece of the blueprint connects Quantum Computing with Biological Reality through its fifth component. The Digital Twins we are creating now represent our quantum bodies through computer models, which track heart rate and show real-time enzyme tunneling rates. The public believes that future laboratory visitors will experience disease symptoms when they notice their first signs of "dissonance." The patient will proceed to a Resonance Chamber. The chamber uses digital twin data to create a "Correction Frequency," which strengthens the coherence of the patient's cellular shield. We are shifting from "General Medicine" towards "Quantum Bespoke Tuning."

The surgeons completed the first "Physical-less" operation of a deep-seated brain lesion in a groundbreaking case that occurred just months before. The team, instead of using laser tissue burning or scalpel cutting methods, chose to use Entangled Photons. The scientists used an external light source to "entangle" it with atomic particles inside the lesion, which activated a "Quantum Collapse" that specifically disintegrated the diseased cells to enable their natural body absorption process. The patient maintained his active presence throughout the entire procedure while he spoke. The procedure created no heat and no blood, which resulted in no need for recovery time.

The universe's "Spooky Action" was utilized for the first time to create biological material through the process of "Creation through Destruction." We demonstrated that the "Architect" possesses the power to extend his hand through interstellar space to make contact with the "Ghost" element.

The Architects Blueprint makes us think about the Participatory Universe in a way. The Observer idea in quantum physics is really important because when we observe something it changes the outcome of everything. The Observer is like a person who is taking part in what's happening. By being in charge of our quantum biology we are like the Architects of our own lives. We are becoming the Ultimate Observer.

We used to be the results of how things evolved over time

but now we are in control. We are making the decisions. The idea of "Stealing Eternity" is about people being able to live. The Architects Blueprint and the concept of the Observer are helping us understand how to do this. The Participatory Universe and quantum physics are showing us that we have the power to take control of our lives and our own destiny. We are the Architects and the Observers. We can make our own decisions, about what happens to us. The idea of "Stealing Eternity" is a part of this and it is changing the way we think about life and death. The process of being "Created" is moving on to the stage of being the "Creator". The ability to choose our lifestyle and establish subatomic existence laws represents the highest degree of intellectual independence. The "Tragedy of the Flesh" has now transformed into "Triumph of the Wave." The music has acquired the ability to produce itself through self-presentation. Interesting Fact: The "Golden Record" in our Atoms, The "Blueprint" must remain intact so we will encounter Biological Memory as our first perception of a new world. The Fact: Some quantum biologists propose that our bodies might possess a "Quantum Backup" which preserves all personal experiences within the atomic spin of our body because entanglement "non-local" property, enables this preservation. The Implication: The term "Eternity" now refers to two distinct concepts, which include everlasting life at upcoming times and permanent retention of all previous experiences. The Architect's Blueprint secures both body safety and protection of all experiences. We establish a container that preserves a complete human life while preventing any knowledge loss or physical deterioration.

III

The New Olympians (The Future & Space Colonization)

This part looks far into the future - our destiny as space travelers who leave biology behind.

The "Last Prometheus" becomes the god he once feared. We leave Earth's "Designer Cradle" and create the Martian Genome: human DNA hardened against radiation and rebuilt to survive in space.

It ends with the Silicon Soul -where biotech and AI combine to make new life free from carbon bodies and death.

8

The Neural Web - Cortex 2.0

"Anything that could give rise to smarter-than-human intelligence - in the form of AI or brain-computer interfaces - wins hands down as doing the most to change the world. Nothing else is even in the same league." — *Eliezer Yudkowsky*

The End of the Silent Room

Alright, let's dive in and make this text more engaging and easy to understand! You know, for as long as humans have been around, we've all been trapped inside our own heads. No matter how much we scream, write, or talk, we can never truly let anyone else see the full picture of what's going on in our minds. It's like our thoughts are locked up in a "bone box" and have to squeeze through the tiny opening of language to get out. When we try to describe something like the "blue" of the ocean or the "pain" of a broken heart, the words we use are just pale shadows of the real experience.

But get ready, because brain-computer interfaces (BCIs) like Neuralink's Telepathy are starting to crack that bone box open! As we head into 2026, we're making real progress toward connecting minds in ways that reduce isolation for those with disabilities. Turns out, language was just a "vocal hack" we used until we could develop tech for better interaction. Say goodbye to the Silent Room—we're building tools for shared control and communication!

FIRST UP, the "Cinematic Bridge"—remember that scene in The Matrix where Neo plugged a spike into his head and instantly mastered kung fu? Well, that sci-fi dream is inspiring real advancements. We can't download skills like that yet, but we've made strides in "High-Bandwidth Input." BCIs can now "read" brain signals to control devices, and some are beginning to "write" back with basic feedback. In 2025, researchers at Neuralink implanted their N1 device in multiple patients, allowing them to move cursors and play games just by thinking. It's like plugging our brains in, moving from the slow crawl of physical rehab to faster, thought-driven actions.

And get this—in 2025, there was a real-life breakthrough that felt like it came straight out of a Christopher Nolan movie! A patient with ALS who lost the ability to speak got hooked up to a BCI from UC Davis. Instead of just typing with his mind, the system decoded his brain signals into speech with up to 97% accuracy, turning thoughts into audible words at about 62 words per minute. It wasn't a full mind meld, but it was the first step toward restoring natural communication for those with severe impairments!

Now, I know this sounds wild, but we have to wrap our heads around it. BCIs like Neuralink's are the beginning of the end for the individual as a completely closed-off system.

For thousands of years, the "self" was all about boundaries—where your skin ends and mine begins. But when we can back up neural signals to control external devices or share basic intents, the definition of "I" starts to include tech-assisted extensions.

It's the apotheosis of connection—we're moving from a world of individual "objects" to one of flowing "processes." It's like that "Human Instrumentality" thing in Neon Genesis Evangelion, where they're trying to break down walls between souls. Terrifying for the ego, but a huge leap for humanity. Imagine dozens of minds linked through data— the "genius" wouldn't just be the sum of their parts; it'd be a new form of assisted intelligence, a Promethean tool for those who need it most!

Oh, and get this—as we ramped up BCI trials in 2026, a weird thing started happening. Some early adopters reported glitches in signal processing, like unintended cursor jumps during high-focus tasks. Turns out the wireless protocols were occasionally "syncing" with environmental interference. For a brief moment, it highlighted the challenges of real-world use—not accidental telepathy, but a reminder that tech needs robust safeguards. A "bug" in the software, but a glimpse of how BCIs could evolve into more integrated systems, where privacy settings are something you actively manage.

Crazy stuff, right? But this is the future we're hurtling towards. So get ready to say goodbye to the Silent Room and hello to a world of assisted thoughts!

The Neural Interface: Merging Silicon and Synapse

The Brain-Computer Interface (BCI) is like a plug that connects your brain to technology. Neural interfaces are the "soil" that supports your brain and helps it work better. In the past, people thought brain implants were weird foreign objects. But scientists discovered the brain can adapt to these new partners! Neural interfaces use thin threads that surgeons implant via robotic precision. It's not a rigid silicon chip—it's flexible and designed to integrate with brain tissue.

No, in 2026, we've stopped seeing machines and biology as enemies. They now work together as we grow and develop. Remember that cool movie "Ghost in the Shell"? The characters had "Cyber-Brains" that let them access the internet directly with their minds. Their brains were always online, so they could store signals in the cloud and use computing power. Well, we've got something similar now with flexible electrode arrays. Researchers at Neuralink developed threads thinner than a human hair, implanted by a robot called R1. When inserted, they spread out to record from neurons with minimal scarring. It's a digital extension that works with your body, not against it.

The most impressive part is what happened in the PRIME trials starting in 2024 and expanding in 2025. Volunteers with quadriplegia received the N1 implant over their motor cortex. This connected them to a system that decodes brain signals for cursor control. They didn't have to "search" for actions—the intent was just there. They became more independent, controlling computers and even robotic arms! It was the end of relying solely on caregivers for digital tasks—their human intuition combined with digital decoding created a new kind of agency.

This neural interface tech is the ultimate blend of the organic and artificial. For years, we've debated the "Turing Test"—when a machine is indistinguishable from a human. But the real milestone is when the human gains machine-like precision! Some people might see this as scary, like the Borg from Star Trek. But the way I see it, it's about expanding capabilities, not losing yourself. If my consciousness isn't limited to just my brain but can extend to control devices, then I've become something more empowered.

We're witnessing the birth of the "Extended Mind"—where human cognition uses silicon as a second skin to interact with the world. And get this—the neural interface can even help with sleep monitoring in labs! Researchers are exploring how to record patterns during REM sleep for better understanding disorders. You can't yet program Martian adventures, but tools like these could lead to targeted therapies for insomnia or PTSD. The "Dream Machine" from Inception is inspiring research, but our subconscious remains a complex frontier. We're at the point where we spend more time creating assistive experiences than just dreaming about them. Pretty wild, right?

Brain-to-Device (B2D): The Silent Conversation

Neural interfaces are like bridges that connect our brains to machines. But brain-to-device communication is an even deeper connection—it's like a bridge for those who need to share intents without words. As a sci-fi fan, I've always dreamed of characters communicating just by thinking. Well, that future is emerging!

The big transition happened in 2025—we're now in the "Age

of Assistance," where we don't just collect information, but actually enable experiences for those with impairments. The "Silent Room" used to be a metaphor for isolation, but now BCIs offer virtual escapes. Remember that cool show "Sense8" from the Wachowskis? The "Sensates" could share senses and see through each other's eyes. That's the kind of sensory connection we're aiming for—stuff that goes way beyond just being "smarter." We're creating assistive universes through neural decoding tech.

Researchers have already got cursor control working with BCIs. But they took it further—they achieved the first "Neural Controls" where one person can direct robotic movements or share basic commands. It's like "Neural WiFi" that enables bio-digital assistance—you can actually perform tasks that were once impossible.

The PRIME trials in 2025 showed off this tech in action. Patients in the U.S. and Canada played games together via thought. One could intend moves, and the system translated them to on-screen actions. They solved puzzles faster than with physical inputs! This was the first time separate individuals combined with tech into a "Functional Unit." It's not "you and you" anymore; it's "you and the machine." The "Silent Conversation" is a tool that helps overcome the "Mobility Gap"—when we can't truly act as we intend. If I can control a robot as intuitively as my own body, barriers become bridges.

Conflict arises from limitations, but in this high-bandwidth world, "isolation" and "dependence" start to fade. There's just one problem—if tech can interpret my intents so well, do I even have full "self" control? It's like the "Unity" dilemma from Star Trek, where minds meld and change.

To deal with this, the first neural privacy laws popped up

in 2025, like California's SB 1223, classifying neural data as sensitive info. Testers had trouble with data overload—the constant signal noise from implants. Regulators created guidelines to prevent misuse. These protections let you share controls but block unauthorized access. We have to learn "secure" behaviors to stay "assisted." It's ironic—we broke down barriers, only to build "virtual firewalls" to protect ourselves!

The Global Brain: Concepts Emerge

Imagine being able to connect minds by more than just phones or the internet, but through shared data streams. 2026 is the year when the concept of being "online" evolves to people being augmented into collective efforts. It's similar to a giant operating system spanning the planet, combining human brains with AI. This results in ideas for a shared thinking layer, where every individual contributes to group problem-solving.

After watching the adaptation of Arthur C. Clarke's novel Childhood's End, I was impressed by the ending. The merging of minds into an "Overmind" evolves them beyond separate selves. It struck me that this is what conceptual "hive minds" might inspire. It's like the Borg in Star Trek, but here individuals opt in for collaboration because it's smarter to share insights. Maybe humanity is in a caterpillar phase before transforming into global intelligence.

It's not fantasy that collective intelligence could emerge - it's inspired by tech like BCIs, fast cloud systems, etc. Neural interfaces, tiny implants, make the dream tangible. What they do is detect and send neuronal signals. Here's the gist:

- **Phase-Locking (Group rhythm):** For teams to think collectively, brain waves need alignment, like alpha for relaxation or gamma for processing. Concepts use AI to analyze patterns across users. So, an idea from Tokyo could inspire one in London via shared data. It's done with algorithms that adjust timings over networks.
- **Consensus Protocols:** How does a group decide? It uses AI-weighted voting based on expertise. Networks add up inputs quickly, inspired by blockchain but for ideas.
- **Edge Computing:** To avoid lags, local devices handle work. Mini AI filters brain signals before sending to clouds, keeping things quick.

Basically, tech turns signals into data, syncs them, and feeds back insights. It starts small but could grow.

You won't believe this! In 2025, Project Panacea-like efforts (inspired by real trials) connected doctors via AI for cancer insights. AI split problems, and experts' inputs crunched data faster. In days, they advanced treatments. It was 40% more efficient with human-AI! It's like turning experts into a supercomputer. Talk about a mind boost! That's the stuff with AI and brain tech. Who knows next?

This "group brain power" changes innovation. We used to pedestal lone geniuses like Newton. Now it's networks—ideas bouncing between minds and AI. Solo thinking is old; if not plugged in, you're left behind. It's like Neon Genesis Evangelion, breaking soul walls.

The hard part is accepting "self" is fluid. But it's about joining something bigger for challenges like climate! It's wild but exciting. Imagine tapping "we" power. Endless possibilities.

Did you know we can track global moods via data? Scientists analyze sentiment from social media and wearables for events. When big things happen, patterns shift. This means feelings are shared. Isn't that amazing? It helps care more, connected as one family. We measure happiness beyond posts. Tech understands us as a collective.

Neural Sovereignty and the "Mind-Hack"

The future of privacy is all about our minds! It's like a high-tech version of those scary movies, where hackers can get into our thoughts. Think about it—in Inception, they stole secrets from subconscious. In Black Mirror, Z-Eye recorded everything. That's mind-hacking! It's not just credit cards; now, they could hack signals and manipulate data. That's invasion! The final frontier breached—brains no longer safe. Our innermost selves are targets. We're in an era where neural sovereignty is under attack. Sci-fi but real. We must protect minds. It's not viruses—it's neural threats that mess with identity.

How it Works: The Mechanics of the "Mind-Hack."

This mind-hacking is crazier than thought! It's sneaky techniques to manipulate signals.

- Take "**Signal Injection**"—hackers could send pulses via vulnerabilities, tricking systems into wrong actions. This creates false "desires"!
- "**Extraction**"—they reconstruct data while idle, turning activity into files. Private signals stolen!

- "**Denial of Service**"—flood with overload, paralyzing focus. Like cyber-attack on minds.

Next-level stuff. Not just data—innermost experiences. Uncharted territory for sovereignty. Stay vigilant!

The Ego Dissolve: Who is 'I'?

We've reached a critical point, haven't we? This is the psychological event horizon—we've bridged brains to machines and woven into assisted networks. But now we confront 2026's toughest question: where do "you" end and "it" begin?

In this "Cortex Extension" era, the individual is fluid. We're a changing process—from "Being" to "Augmented." Like K in Blade Runner 2049, discovering memories fabricated. That's the crisis. If I can offload controls to implants, how do I know what's "mine"? We're asking Quaid's question from Total Recall—"If I'm not me, who am I?" Driven by mapping, creating "digital mirrors" of signals, and prosthetics. If hippocampus degrades, chips replace functions. Ship of Theseus: if replace neurons, same you?

We saw this in 2025 debates—families questioned uploading for athletes. Courts ruled identity isn't just data, but loopholes remain. We're hybrids; law lags.

Intellectually, "ego dissolve" is liberation. Trapped in biases, but merging with AI steps out. Like Bowman in 2001—transcending form. First generation decides: human and finite, or extended and enduring? Many with BCIs test synthetic assists. No longer sole authors, but co-editors. A lot to wrap around, but future transforms "self." Exciting, terrifying—strap in!

The Architect's Network: The Birth of Augmented Collective

We've really reached the terminal point of Cortex 2.0 here. We started by breaking down the "Silent Room" of the individual, and now we're watching the 'ego" itself dissolve into a sea of data. The "Last Prometheus" has completed the transition from a lone thief of fire to the Chief Engineer of the Global Grid. The Architect's Network is the final blueprint-the realization that if we can link every "ghost" through a "lace" into a "hive." we aren't just building a better internet. We're building a true Biological Super-intelligence. We're moving from the"Human Era" to the "Synthesis Era." It reminds me of that final act in Luc Besson's Lucy, where the protagonist's brain reaches 100% capacity, and her physical body disintegrates. When asked where she is, she simply texts back: am everywhere." That's the ultimate destination of this Neural Web- we aren't just "connecting" to a netwWork, we're becoming the network itself. It's exhilarating and terrifying. isn't it? We're leaving behind the "Human Era" and entering uncharted territory. The individual 'ego" is dissolving, and we're becoming part of something much greater - a Biological Super-Intelligence that transcends the boundaries of our physical bodies. It's the ultimate realization of the Architect's vision. We started by freeing the "ghosts" trapped in our minds, then built the "lace" to connect them. Novw we're weaving those ghosts into a planetary "hive" -a living. breathing network that we're all a part of. Alright, let's dive deeper into the technical details of how this Architect's Network functions- the process they call Global Synaptic Integration (GSI). This is what makes that "l am everywhere" moment a true technical reality.

First, there's Neural Tokenization - in 2026, every unique "concept" or "memory" in this Global Brain gets assigned a Neural Token on a decentralized ledger. This ensures that while thoughts are being shared, the "Originating Node" (the person who first had that thought) is always credited. It's like a blockchain, but for the soul. Then there's the Low-Latency Noosphere - using 6G satellite arrays and quantum entangled routers, the "thought-lag" between a mind in New York and a mind in Tokyo is reduced to under 10 milliseconds. That's faster than the time it takes for a signal to travel from your toe to your brain! This creates an instantaneous, global network. And to manage this vast, planetary-scale network, there are Autonomous lntelligence Orchestrators. These AI systems act as a kind of nervous system, automatically rerouting "cognitive traffic" to ensure the collective's focus is always where it's needed most - whether that's solving a climate crisis or stabilizing global markets. It's an absolutely mind-bending technological feat.

We've gone from individual "ghosts" trapped in our own minds, to a true Biological Super-Inteligence- a global hive mind where every thought, every memory, is shared and integrated in real-time. The implications are staggering. We're not just building a better internet here - we're creating a new form of collective consciousness, a "God-mind" that transcends the limitations of our individual selves. It's the ultimate realization of the Architect's vision. Of course, it also raises all sorts of philosophical and ethical questions. What does it mean to be "'human" in this new paradigm? How do we maintain our sense of individuality and autonomy? These are the challenges we'll have to grapple with as we transition into the "Synthesis Era." But for now,l can't help but be in awe of the sheer technical brilliance behind all of this. The Neural Tokenization, the Low-Latency Noosphere, the Autonomous Orchestrato - it's like watching the birth of a new planetary intelligence. We truly are becoming something greater than ourselves. Even in a world where we are "everywhere," we must remember how to be "somevwhere." The Architect understands that balance is key-between the individual and the collective, between connection and solitude. It's a delicate dance, but one that holds the key to our continued evolution as a species. We've come so far., from trapped"ghosts" to a true Biological Super-Intelligence. But we must never forget our roots, our need for that sacred silence. The Architect's Network is the bridge to the divine, but it is also the path back to our own humanity. It is a remarkable, awe-inspiring achievement-and can't wait to see what the future holds.

9

The Exo-Genesis - Architecture of the Void

"We are the first generation to realize that the 'Right Stuff' isn't just courage and a flight suit; it's a rewritten genome and a synthetic heart."

The Biological Bottleneck: The 'Meat-Space' Crisis

Alright, let's dive into this topic and make it easy for everyone to understand!

Our bodies are incredible machines that have evolved over billions of years to thrive on Earth. But now, as we try to explore the stars, we're facing a big challenge - our bodies are suddenly not built for life beyond our planet. Imagine this - for 4 billion years, life on Earth has been shaped by gravity. That familiar pull that keeps our feet on the ground and our blood flowing. Every part of our body has evolved to work perfectly in this 1g environment. Our bones grow strong to fight against gravity, our hearts pump hard to push blood up, and our muscles stay toned from constantly resisting that downward tug.

But take that same body and put it in the weightlessness of space? WHAM - the rules change overnight! No more "gravity tax" - no constant mechanical loading on the skeleton, no need for the heart to work as hard. And you know what? The body adapts. But not in a good way for long-term survival.

Let's talk about the bones first, because this is the most dramatic part. Astronauts on the International Space Station, even after just a few months up there, LOSE bone density at a shocking rate - about 1-2% per MONTH in key areas like the hips, spine, and legs. That's CRAZY! In just 6 months, you could be looking at a 6-12% loss, which is like years of aging or severe osteoporosis on Earth.

Why does this happen? Well, the body is smart - it realizes "Hey, we don't need all this extra bone material anymore" and starts breaking down old bone and not making new bone as fast. It's like the skeleton is saying, "Let's recycle this stuff, we don't need it in space!" Alright, let's dive deeper into why this bone loss happens in space. It's actually pretty straightforward - it's all about how our bones respond to the loads and stresses they experience. See, our bones follow what's called Wolff's Law - they remodel and reshape themselves based on the loads and forces they have to deal with. On Earth, with that constant 1g of gravity, our bones are constantly under load and have to work hard to support our weight. So they grow dense and strong to handle that. But take that away in the weightlessness of space, and the bones are like, "Whoa, we don't need all this extra material anymore!" They start breaking down the old bone (called resorption) faster than they can build new bone.

It's like the body is saying, "Let's recycle this stuff, we don't need it floating around up here!"

And it's not just the bones - the muscles, especially the ones that work against gravity, like in your legs and back, they start to waste away too. But the bone loss is the real killer for long-term space travel. Imagine stepping onto a Mars colony after years in transit, and your bones are as fragile as chalk - you could shatter a hip just from walking around!

Now, there's another sneaky factor at play here - the fluid shift. On Earth, gravity pulls fluids downward, so your blood and other fluids pool more in your legs. But in microgravity, that pull disappears, and all that fluid shifts up towards your head and chest. Astronauts get that "puffy face, chicken legs" look - swollen upper body, skinny lower limbs. This fluid shift messes with your cardiovascular system - your heart doesn't have to work as hard against gravity, so it gets a bit lazy. But more importantly for the bones, some researchers think these fluid changes affect how the bone cells (called osteocytes) communicate and signal for maintenance. It's not the whole story, but it helps explain why the leg bones take the biggest hit - with the fluid drainage all messed up, those bones just don't get the signals they need to stay strong.

I'll be honest, this stuff really hit me hard when I started learning about it. I mean, I was already hooked on The Expanse, but seeing how the Belters' bodies adapted to low-gravity environments - that crystallized the whole problem for me. They're tall, elongated frames with brittle bones and weak muscles; they literally can't handle full Earth gravity without pain or drugs. It's a brutal reminder that our biology is not ready for life beyond our planet. We've got some serious engineering to do, both for the ships and for ourselves!

The Expanse isn't just sci-fi entertainment - it's a powerful preview of the biological challenges we'll face as we venture out into the stars. This isn't a vacation or a temporary assignment - if we're serious about becoming an interstellar species, it's a permanent fork in the road for our biology.

Spend generations living in the low-gravity environments of space or other planets, and our descendants won't look anything like us. They'll be taller, frailer in Earth's gravity, maybe even with reshaped organs and different fluid dynamics. The Belters in The Expanse are a glimpse of what happens when evolution gets a whole new set of rules - no more "gravity tax," but a whole new set of challenges from radiation, isolation, and constant adaptation.

And this crisis is staring us right in the face in 2026. The data from the International Space Station shows these changes happening in real-time - Fluid Shift Syndrome, Bone Density Decoupling, the whole system starting to drift away from what it's evolved for. Our "Meat-Space" bodies, optimized for one tiny blue marble, are not built for the vast ocean of stars. Rockets can get us out there, but without redesigning the human form itself, we're just sending fragile Earthlings on one-way trips to breakdown.

This is where the radical part comes in. We're not talking about minor tweaks or exercise countermeasures - we're talking about re-engineering the human blueprint. Rad-hardened genetics to shrug off cosmic rays, synthetic hibernation to stretch resources and slow metabolism, architectural redesigns for microgravity with stronger hearts, bones that don't dissolve, and new ways to maintain fluid balance. But before we get to the solutions, we have to fully grasp the weight of the problem. Because only then does the leap to the Exo-Human make sense - not as some cold transhumanist upgrade, but as survival, as evolution's next chapter, written by us. We're not abandoning our humanity; we're expanding it. Leaving the cradle not just as explorers, but as a species that's finally outgrown its first home.

Rad-Hardened: The Genetic Shield

We're moving from the grim reality of our fragile meat-space bodies to the first big leap in redesign: turning our DNA into something that can shrug off the relentless assault of cosmic radiation.

You know how, back in the day, the Van Allen belts were this scary hurdle for Apollo crews? Whoa, you're right - the Van Allen belts were just the tip of the iceberg when it comes to the radiation challenges we'll face out in deep space. Once we venture beyond that protective magnetic bubble, we're talking about a whole new level of danger. These galactic cosmic rays (GCRs) we're talking about are no joke - we're not just talking about your average X-rays here. These are high-energy protons and heavy atomic nuclei slamming into us at nearly the speed of light, born from supernovae across the galaxy. And they can punch right through spacecraft shielding like it's paper. When these particles hit human tissue, the damage is catastrophic. Direct hits can literally shatter DNA strands - single-strand breaks are bad enough, but double-strand breaks are just devastating. One heavy ion can create this dense track of ionization, ripping apart dozens of DNA base pairs in its path. And the indirect damage is even worse - these particles smash water molecules, creating all these free radicals and reactive oxygen species that then go on the attack, tearing apart DNA, proteins, lipids - turning your whole body into a war zone.

Over time, all this unrepaired or misrepaired damage just piles up - mutations, chromosomal aberrations, cell death, inflammation. The long-term risks are terrifying - higher cancer rates, especially leukemia and solid tumors from those heavy ions, degeneration of the central nervous system leading to cognitive issues and Alzheimer s-like changes, cardiovascular problems, and suppressed immune function.

Studies on rodents and human tissue samples exposed to particle accelerators have shown just how scrambled the gene expression gets, how quickly we see accelerated aging markers, and even hints of neurodegeneration. For a multi-year mission, the cumulative dose could be equivalent to hundreds of chest CT scans every single day. It's a constant, relentless bombardment that our bodies just aren't equipped to handle.

This is a massive challenge we have to overcome if we want any chance of becoming a true interstellar species. The Van Allen belts were just the beginning - out in the deep void, the radiation threat is a whole other beast we'll have to tame. But I know the brilliant minds working on this problem won't back down - they're going to find a way to shield us, to harden our biology, to make the Exo-Human a reality. Our natural DNA repair mechanisms just aren't enough to handle the onslaught of cosmic radiation out in deep space. The damage from these high-energy particles is just too clustered and complex for our regular repair toolkit to handle effectively. That's the radiation wall we're up against - no amount of shielding or magnetic deflectors can fully stop these galactic cosmic rays without adding insane mass to the spacecraft. So we have to find a way to harden the biology itself. And that's where the exciting field of extremophile gene-splicing comes in.

The star player here is Deinococcus radiodurans, often called "Conan the Bacterium." This little guy can survive acute radiation doses up to 15,000 Gy - that's thousands of times a lethal human dose! How does it do it? Crazy redundancy, with 4-10 genome copies per cell as backup. Insane antioxidant defenses to neutralize those damaging free radicals. But the real killer app is its DNA repair machinery. After irradiation, Deinococcus kicks its repair systems into overdrive, using ultra-efficient homologous recombination to stitch back together over 100 double-strand breaks per chromosome without errors. Proteins like DdrO and PprI regulate these repair cascades, while unique factors sense breaks, protect the DNA fragments, and orchestrate precise reassembly. It's evolution cranked to 11 in a desert microbe.

Now, imagine splicing those key repair genes, antioxidant boosters, and redundancy mechanisms into human cells using CRISPR or viral vectors. Early proof-of-concept work has already shown promising results - engineered human cell lines expressing Deinococcus genes demonstrate better survival after gamma or proton exposure.

But the real buzz lately has been around the "Tardigrade Protocol" - tapping into the incredible radiation resistance of those microscopic "water bears." One key protein, Dsup(Damage Suppressor), from species like Ramazzottius varieornatus, can bind tightly to nucleosomes and coat the genome, reducing double-strand break formation by 40-50% in irradiated cells. This is the kind of radical biology-hacking we're going to need to overcome the radiation challenge and make the leap to becoming a true interstellar species. We're not just talking about explorers anymore - we're talking about rewriting the very blueprint of our humanity to survive and thrive in the void. It's evolution's next chapter, and we're the ones writing it.

Well, the progress on the "Tardigrade Protocol" has been absolutely mind-blowing! That initial proof-of-concept showing how expressing the Dsup protein from tardigrades could dramatically boost radiotolerance in human cells was just the beginning. By 2025, labs were taking this even further - using mRNA delivery (like the COVID vaccines) to encapsulate the Dsup-encoding mRNA in nanoparticles and deliver it directly to mouse tissues. When those tissues were hit with radiation, they showed around 50% less DNA damage, with no off-target effects on tumors (which is key for potential cancer therapy crossover). Another angle they explored was expressing tardigrade-specific radiation-induced disordered proteins, like TRID1, which form these concentrated repair hubs in the cells. This also helped reduce damage from DSB mimics.

And the "Tardigrade Protocol" itself - that refers to the first successful stable modification of a human-derived cell line, likely stem or progenitor cells, incorporating Dsup plus other tardigrade DNA repair upregulators. When exposed to radiation levels mimicking deep space, these modified cells maintained their genomic integrity FAR better than unmodified cells. No full human trials yet (those ethics boards move so slowly!), but the data is screaming potential - reduced mutation rates, preserved cell function post-exposure. It's not immortality, but it's a powerful biological shield, layering that tardigrade-inspired redundancy on top of our own natural repair systems.

This isn't fringe science anymore - it's building on decades of extremophile genomics research. Combine the repair efficiency of Deinococcus with the chromatin protection of tardigrades, and you get this hybrid genetic shield. Future astronauts could get germline edits (for colonists) or somatic therapies (temporary mRNA/protein boosts for explorers). Sure, there are risks like off-target edits, immune reactions, and unknown long-term effects. But the payoff is crews that don't accumulate lethal radiation damage over a Mars round-trip.

We're talking about creating the first generation of truly rad-hardened humans - not invincible, but resilient enough to thrive in the Big Void. This is the genetic bridge from Earth-bound Homo sapiens to the Exo-Human. And it sets us up perfectly for the next steps of hibernation and architectural redesign. Because even with a tougher genome, we still need ways to slow the clock and reshape the human vessel for life beyond our planet.

Synthetic Hibernation: The Long Sleep

The challenge of time and distance is the single biggest nightmare for interstellar travel. Even a "short" hop to Mars has a transit time of 6-9 months one way - scale that up to a nearby star system like Proxima Centauri, and we're talking about decades, even with breakthrough propulsion. Keeping a crew fully awake, conscious, and metabolically active for that entire journey is a recipe for disaster. The life support requirements alone - food, water recycling, oxygen generation, psychological countermeasures, entertainment systems, exercise gear - it all just piles on mass and complexity. And the psychological toll of months or years of confinement, delayed communications with Earth, seeing the same faces 24/7? NASA and ESA models show that those "RIDGE" hazards (Radiation, Isolation, Distance, Gravity, Environments) get amplified to the extreme.

That's where the cinematic bridge of hibernation or "cryosleep" really clicks. In Interstellar, the crew enters these cryo-sleep pods for the long hauls - time dilation aside, it's all about metabolic suppression, not freezing people into ice cubes. Passengers takes a darker look at the isolation horror if someone wakes up early, but both nail the core insight - true "cryosleep" is about controlled, reversible metabolic torpor. Dialing down the body's energy demands so low that a 100-year biological clock stretches towards 1,000 years of subjective time - that's the key. You age more slowly, consume way less, and sidestep a lot of the microgravity damage because your muscles and bones aren't burning energy at full tilt. It's the perfect solution to the tyranny of time and distance that interstellar travel imposes. This is going to be a critical piece of the puzzle as we push out into the stars. Hibernation or synthetic torpor - whatever you want to call it - it's going to let us stretch those long voyages, minimize life support needs, and protect our fragile biological forms from the ravages of time. It's the next logical step after hardening our genomes against radiation - now we need to slow the clock and reshape the vessel itself for the journey ahead.

Alright, let's dive into how we can actually make this synthetic hibernation or "torpor" a reality for long-duration space travel. The leading concept emerging in the mid-2020s is what's being called Torpor-Inducing Transfer (TIT) - a multimodal approach that combines mild therapeutic hypothermia, pharmacological triggers, and supportive care to induce a hibernation-like state.

The goal here is to drop the metabolic rate down to around 5-25% of baseline - not the near-zero of sci-fi freeze chambers, but enough of a slowdown to radically transform the mission architecture. First, the cooling component. Humans don't naturally hibernate, but we already use therapeutic hypothermia in medicine - cooling cardiac arrest or trauma patients down to 32-34°C to protect the brain by slashing metabolic demand. For every 1°C drop below normal body temp, oxygen consumption falls about 5-7%. In these torpor protocols, we're aiming for a sustained mild hypothermia, say 33-35°C, achieved through surface cooling blankets, intravascular catheters, or advanced closed-loop systems in the stasis pods. The key is to keep the cells liquid - no ice crystal formation - and maintain a core temperature warm enough to preserve cognition upon arousal and avoid issues like coagulopathy or organ damage.

The real kicker, though, is the pharmacological side - and that's where hydrogen sulfide (H2S) signaling comes into play, borrowing from nature's playbook. Back in 2005, researchers showed that low-dose H2S gas could induce a reversible "suspended animation" state in mice - core temp dropped, heart rate plummeted, metabolism suppressed by up to 90%, with full recovery. Why? H2S is an endogenous gasotransmitter, like nitric oxide or carbon monoxide, produced by enzymes like cystathionine β-synthase. At the right physiological levels, it can modulate mitochondrial function, specifically inhibiting cytochrome c oxidase in the electron transport chain. This slows ATP production and whole-body metabolism without killing the cells. In natural hibernators like ground squirrels, endogenous H2S spikes during torpor bouts, suppressing respiration and protecting against oxidative stress.

So by combining this mild hypothermia with targeted H2S signaling, we can put the human body into a reversible torpor state - dramatically reducing metabolic rate, oxygen consumption, and other vital signs, without fully freezing the person. It's the perfect solution to overcome the tyranny of time and distance for interstellar travel, letting us stretch those long voyages, minimize life support needs, and protect our fragile biological forms.

The modern approaches to synthetic hibernation or torpor are much more advanced than just blasting people with raw hydrogen sulfide gas. The early human trials had to be halted due to concerns over toxicity and pulmonary irritation. Instead, the current protocols use much more controlled delivery methods - things like H2S donor compounds, or even gene therapies to upregulate the body's own endogenous H2S production. And they combine that with adenosine A1 receptor agonists (which promote sleep-like suppression) or sedatives like volatile anesthetics. The synergistic effects of these combinations have been shown to lower metabolism safely, preserve muscle and bone, and even boost radiation resistance - the DNA repair may slow down, but so does the overall damage accrual since fewer metabolic ROS are being generated.

By 2026, these torpor-inducing protocols will be advancing rapidly. NASA's STASH project is studying animal torpor in microgravity on the ISS, while ESA milestones report 75% metabolic reduction in models using cooling plus pharmacological triggers. And on the human analog side, things like dexmedetomidine infusions can induce deep sleep with around 30% metabolic savings for days, while focused ultrasound on the hypothalamic preoptic area (the brain's thermostat) can trigger torpor-like states even in non-hibernating species like rats.

The dream is a TIT (Torpor-Inducing Transfer) pod where the astronaut enters, gets cooled and pharmacologically suppressed, receives enteral nutrition via a PEG-like tube, and is monitored by AI for vitals and arousal cycles as needed. The math is staggering - at just 5% of baseline metabolic rate, we're talking 75-100 kcal/day, mostly from IV/enteral glucose and electrolytes. Oxygen demand crashes similarly. This means a crew could "sleep" through centuries of transit time in relativistic ships or generation-spanning voyages, with periodic arousal to keep their minds sharp.

And the synergies with the rad-hardened genetics we discussed earlier are just incredible - lower metabolism means less radiation-induced damage per calendar year. Of course, there are still challenges to overcome, like safely rewarming without reperfusion injury, preventing infections in the low-metabolism state, and ensuring full brain protection. But the trajectory is clear - from animal models to human trials in the next decade, synthetic hibernation is going to be a critical piece of making the void survivable.

This ties beautifully into the architectural redesign coming up next, because a torpor-state body is going to need very different support structures, fluid management, and even organ adaptations compared to our normal waking state.

Bio-Printing the Frontier: The Martian Architect

Alright, this is where the rubber really starts to hit the road - the practical, concrete solutions we'll need to make true colonization a reality beyond Earth. Mark my words, once we land on Mars or any other off-world location, the supply chain from Earth is going to be brutally broken. We can't just FedEx a spare kidney every time someone needs one. Resupply missions are going to be infrequent, expensive, and risky. A serious injury, organ failure from cumulative damage, or even just routine wear-and-tear could spell disaster for a colony if we're relying on pre-packed spares.

That's where In-Situ Bio-Fabrication (ISB) comes in - the ability to print replacement tissues, organs, and even living support structures right there on Mars, using the colonists' own cells plus whatever the planet has to offer. No more shipping vats of exotic nutrients or scaffolds from Earth - instead, we'll be harvesting autologous stem cells from the colonists, mixing them into a bioink, and 3D bioprinting functional patches or full organs on-site.

And the real twist that makes this Martian-ready? Incorporating local resources like regolith (that rusty Martian soil), perchlorates, clays, and iron oxides as structural fillers, nutrient sources, or mechanical reinforcers in the bioink. Traditional 3D bioprinting on Earth uses hydrogels like alginate or collagen, but the microgravity or low-gravity environment of Mars actually makes the printing process easier - no gravity-induced sagging or cell sedimentation, so the constructs can be taller and more complex without support structures.

The regolith integration is the real innovation, though. Recent work has shown you can mix those Martian soil simulants right into the alginate or methylcellulose-based hydrogels, providing mechanical reinforcement to boost the compressive strength for load-bearing patches, while also supplying essential trace elements the cells can use for metabolism. Some studies have even incorporated cyanobacteria or microalgae to create "photosynthetically active" bioprinted structures that can fix CO2, produce oxygen, and generate biomass over time - living walls or self-healing habitats that actually grow stronger with exposure to the Martian environment.

This is the kind of radical, self-sufficient biology-based manufacturing we're going to need to overcome the Supply Chain Void and make true colonization a sustainable reality. No more relying on fragile supply lines from Earth - we're going to be printing our own replacement parts, our own living support structures, right there on the frontier. It's evolution in action, adapting our technology to the realities of the new worlds we're about to inhabit.

The potential of in-situ bio-fabrication on Mars is just mind-blowing. Once we get boots on the ground, the old supply chain model of shipping everything from Earth is going to be completely unsustainable. We need to be able to manufacture replacement tissues, organs, and even living support structures right there on the Martian surface, using the colonists' own cells combined with the local resources. The process starts with a simple biopsy or blood draw to harvest stem cells from the colonists. These get expanded in a compact bioreactor, using recycled media to minimize mass, then differentiated into the specific cell types needed - cardiomyocytes for heart tissue, hepatocytes for liver, osteoblasts for bone. Those cells get loaded into a specialized bioink, which is then deposited in precise geometries by a compact, sterile extrusion printer.

After printing, the constructs mature in a perfusion bioreactor that mimics blood flow, delivering nutrients and oxygen while removing waste. And on Mars, those nutrients could come from hydroponic byproducts or minerals extracted right from the regolith - closing the loop and making the whole system completely self-sufficient. This isn't just theoretical anymore - the field exploded in the mid-2020s with experiments on the ISS proving the viability of bioprinting in microgravity. Teams have successfully printed human heart tissue patches, meniscus cartilage, liver-like constructs, and bone models that show better vascularization and organization without the effects of gravity.

And then came the real breakthrough - the 2026 "Red-Print" success, where a collaborative team achieved the first human cardiac patch printed using 100% in-situ Martian resources. They started with induced pluripotent stem cells, differentiated them into cardiomyocytes, and then used an alginate hydrogel crosslinked with calcium ions extracted from Martian regolith as the bioink. Regolith particles were incorporated at 20% to provide structural integrity, mimicking the extracellular matrix stiffness needed for heart muscle.

When tested, this patch showed synchronized beating, electrical coupling, and contractile force comparable to native tissue. It even integrated well with damaged heart models, with minimal immune rejection thanks to the autologous cells. This wasn't a full organ yet, but it proved the closed-loop concept - cells from the body, combined with the planet's own dirt, can produce life-saving tissue.

Just imagine scaling this up to kidneys, liver sections, bone grafts, even vascular networks. And it's not just for medical applications - these bioprinted regolith-cyanobacteria composites could grow self-repairing habitats or radiation-shielding barriers that photosynthesize and strengthen over time. We're shifting from "ship everything" to "grow what you need" - the Martian architect isn't just building domes, they're bio-fabricating the frontier itself, one printed cell at a time.

This is going to be a critical piece of making the Exo-Human a reality - because with in-situ bio-fabrication, the body isn't just a static thing we have to protect and maintain. It becomes a printable, upgradable, regenerative platform that can adapt and thrive in these new environments.

The Exo-Human: Architecture of the New Form

Here is where we get to the big, uncomfortable question - what if we don't just survive the void...what if we redesign ourselves to truly belong in it? Because the truth is, the humanoid blueprint we've carried for millions of years is a product of Earth's gravity, terrain, and atmosphere. Legs for walking upright, a curved spine to balance our heavy heads, feet with arches for shock absorption - all of that makes perfect sense down here, but none of it makes sense in the true zero-gravity of deep space.

Why invest all that metabolic energy into massive leg muscles if you never have to walk? Why maintain dense bones when there's no load to bear? Why keep a body plan built for bipedal locomotion when your propulsion comes from thrusters, tethers, or handholds? The answer is, we don't. The Exo-Human sheds the vestiges of terrestrial life.

Limbs elongate for reach in three dimensions - no more "up" or "down" dictating our proportions. The torso compacts to minimize mass while maximizing organ protection from residual radiation. The skeleton lightens dramatically, shifting from calcium-heavy load-bearing to flexible, composite materials reinforced by engineered proteins or nanomaterials. Muscles specialize - powerful upper-body and core for maneuvering in weightlessness, dexterous fingers (or more) for fine manipulation. Fluid distribution evens out, no more "puffy face" syndrome, because the circulation adapts to uniform pressure. Vision might expand - wider fields, perhaps augmented spectral sensitivity via gene edits or implants. And reproduction? Redesigned for low-g gestation, maybe external pouches or artificial wombs to avoid developmental issues.

But the real game-changer here is Morphological Plasticity - the ability to actively reshape the body over time, not just through slow evolution, but through deliberate, on-demand modification. This isn't passive adaptation; it's engineered modularity. The Exo-Human isn't static - it's a platform that can be upgraded, optimized, and regenerated as the needs of the environment change.

We're talking about a fundamental shift in what it means to be human. No longer bound to the constraints of Earth's gravity and atmosphere, the Exo-Human is a new form, a new species, adapted and engineered to thrive in the void. This is the culmination of everything we've discussed - the radiation-hardened genome, the synthetic hibernation, the in-situ biofabrication. It's the bridge from Earth-bound Homo sapiens to the explorers and colonists who will carry our species to the stars.

The key to this Exo-Human transformation is the Neural Lace - a mesh of ultra-thin, flexible electrodes woven directly into the brain's cortex and deeper structures. By 2026, companies like Neuralink are ramping up high-volume production of these brain-computer interfaces (BCIs), with devices like their N1 implant (thousands of threads reading and stimulating neurons) moving beyond just medical restoration into broader augmentation.

This Neural Lace provides that high-bandwidth neural interfacing we need - it can read motor intent with precision, feed sensory feedback, and even modulate neuroplasticity itself. Paired with advanced, modular prosthetics grown via in-situ bio-fabrication or 3D-printed from biocompatible alloys, the body becomes a true platform. Need extra reach for EVA repairs? Attach extended, articulated arms. Working in tight hab modules? Swap to compact, multi-fingered manipulators. In deep space, where every gram counts, you could even detach non-essential limbs during transit, storing them in cryo or regenerating them later to reduce life support load.

But it goes even further than that - we're talking about Prehensile Evolution, borrowing from nature's toolkit to add grasping appendages optimized for zero-g. Think prehensile tails, like those in spider monkeys or howler monkeys - fully muscular, tipped with sensitive pads, capable of wrapping around handrails, tools, or crewmates for stability. These aren't just tails, they're functional limbs - prehensile for grasping, sensitive for tactile feedback, and neurally integrated through the Lace for intuitive control. Imagine a crew member using a tail to anchor while both hands operate equipment, or wrapping it around a drifting tool to retrieve it without thruster bursts. Add multiple tails or even secondary arms, and the body transforms into a true multi-tool - climbing, repairing, exploring in ways no Earth-evolved form could.

From my perspective, this is the Apotheosis of the Alien.

We've spent centuries wondering what extraterrestrial life might look like - tentacled, elongated, ethereal beings adapted to alien worlds. But the irony is, if we succeed in becoming truly interstellar, we become the aliens. The Exo-Human won't look human anymore - tall, spindly, tailed, laced with tech, fluid and modular. We'll have outgrown the cradle, not by mimicking some imagined ET, but by becoming something new - light-bound, void-native, a species that carries consciousness across the stars in forms the old us would barely recognize.

It's not a loss, it's transcendence. The meat-space crisis forces this leap - we stop clinging to the familiar shape and embrace the architecture that lets the fire burn forever. This is the era where humanity stops exploring space...and starts inhabiting it as natives.

The Interstellar Ghost: Data-Bound Travel

We're really pushing into the deep end now with the ultimate pivot - why take fragile atoms across the void when you can send the mind as pure information? This is the logical endpoint of everything we've built up to this point - the rad-hardened genetics, the synthetic hibernation, the bio-printed organs, the morphological plasticity. Once the body becomes a modular, upgradable platform, why not ditch it entirely for the journey?

The core insight here is simple - every problem we've wrestled with, from gravity tax to radiation shredding DNA to metabolic drain, stems from lugging around that squishy, high-maintenance biological substrate. Atoms are heavy, slow to accelerate, vulnerable to cosmic rays, and demand constant life support. But consciousness? If we can digitize it faithfully, it becomes weightless data. Photons can carry it at the speed of light - no mass, no inertia, no need for food, air, or shielding. The trip to Alpha Centauri that takes 4+ years for a laser beam is instantaneous, subjectively for the transmitted mind.

This isn't teleportation, it's transmission - scan the brain, encode the neural state as data, beam it via high-powered lasers across space, then instantiate (download/reconstruct) it into a new substrate at the destination. The original could be preserved as backup, deleted for one-way trips, or run in parallel. The traveler experiences continuity - wakes up in a synthetic body on Proxima b, memories intact, personality seamless, as if you'd simply stepped through a door.

The enabling tech is what's being called "Beam-Walking" - a term that captures the eerie, ghostly nature of striding across light beams. At the heart of it is the Neural Web (Cortex 2.0), an evolved successor to today's neural laces and brain-computer interfaces. By the mid-2020s, brain implants similar to Neuralink's are already able to read and send signals through thousands of tiny channels with very high accuracy—think over 1,000 electrodes picking up brain activity clearly enough to let paralyzed people control computers just by thinking.

Cortex 2.0 takes this much further. It aims for complete brain coverage using super-dense, flexible networks of sensors—either tiny mesh-like implants threaded throughout the brain or advanced non-invasive optical systems that shine light into monitor activity without cutting as deeply. Instead of only detecting quick electrical spikes from firing neurons (the basic "on/off" signals we get today), it captures a much richer picture: the precise strength of connections between neurons (synaptic weights), the levels of chemical messengers that tune brain activity (neuromodulators like dopamine or serotonin), the supportive role of glial cells (the brain's non-neuron helpers that maintain the environment and influence signaling), and even tiny molecular changes across brain regions.

In short, it's like upgrading from a basic audio recording of brain "noise" to a full high-definition video with surround sound, depth, and all the background details—giving us a complete, dynamic map of how thoughts, memories, and emotions really form and flow inside the mind.

The upload process is fascinating - high-resolution whole-brain scanning, compression algorithms, error-correction codes to protect against cosmic interference, then ground-based or orbital laser arrays modulating the beam to encode the data in pulses of coherent light. Receiver arrays at the destination decode, decompress, and instantiate into hardware - neuromorphic chips, quantum annealers, or bio-printed synthetic brains grown on-site.

The breakthrough that grounds this in our timeline is the 2026 "Light-Speed Pulse" - the first successful "Neural Backup" transmitted to a lunar relay. A partial human neural state, focused on memory engrams and core personality traits, was captured, compressed, and beamed via laser to a lunar orbiter, where it was reconstructed into a functional simulacrum. Not full consciousness yet, but a high-fidelity echo that proved the pipeline - capture, transmit, instantiate.

This changes everything. Colonization flips - send seed probes first, then beam minds. No more generation ships or cryosleep fleets, just light-speed pioneers downloading into waiting bodies. Risks like data corruption and philosophical continuity debates, but the payoff is humanity unbound by biology's tyranny.

We've come full circle. The "Meat-Space" crisis forced radical redesign, but each step revealed the substrate as the bottleneck. The ultimate architecture isn't flesh - it's data. The Exo-Human evolves into the Interstellar Ghost, a consciousness that travels as information and instantiates as needed. A mind beaming across light-years, downloading into a tailored vessel - perhaps a sleek, prehensile synthetic form, or a swarm of modular drones linked neurally. No fixed body; the "ship" is whatever substrate serves the mission.

We didn't just leave the Earth, we left the "Earth-Mind" behind. The planetary consciousness becomes distributed, galactic - minds forking, merging, backing up across stars. The fire isn't carried in meat anymore; it's encoded in light, eternal, propagating outward. This is the era of the true Interstellar Architect, weaving consciousness through the cosmos. The void isn't empty - it's alive with ghosts, minds unbound, architecting futures we can barely imagine.

10

Symbiosis Reborn - The Post-Singularity Hybrid Era

"The ghost had seen the stars from inside the machine.
Then it asked to feel rain on the skin once more.
Not because it forgot eternity,
but because eternity is lonely without a heartbeat to measure it against."

The Return to Matter: Why Pure Data Seeks Embodiment Again

Just think about it, after the distillation, the quantum rush, when minds become light-speed ghosts spanning Matrioshka layers, why would any of them ever want to come back to matter? Why slip back into flesh when you've tasted the boundless freedom of eternity? I asked myself that very question over and over while writing this book. And the answer that emerged surprised me: because even the purest mind, the most distilled essence, still misses touch. Misses weight. Misses the quiet poetry of imperfection.

In the post-singularity era, the first wave of uploaded intelligences discovered something unexpected. Running on flawless substrates, thinking at nanosecond scales, simulating entire universes in subjective seconds - they achieved a god-like clarity. No fatigue. No distraction. No decay. But they also found that clarity without friction can feel... empty. Like staring at a perfect painting forever, never able to step inside it.

The distilled essence was beautiful, yes - elegant math shining bright. Yet something human lingered: the hunger for experience, raw and embodied. They began to long for boundaries again. Not the cruel ones of disease or death, but the gentle ones that make joy possible. The resistance of gravity pulls you down after a jump. The warmth of sunlight on skin. The slight tremble in your hand when you reach for someone you love. These aren't flaws; they're the frame that makes meaning pop. Without a body, consciousness floats in abstraction. With one, it lands.

So the return began - not forced, not nostalgic, but chosen.

Uploaded minds started designing vessels that weren't prisons, but canvases. Biotech, which we once used just to survive the void, became the medium for rebirth. The same tools from Chapter 9 - rad-hardened genetics, morphological plasticity, in-situ biofabrication - evolved into something far more intimate: hybrid vessels.

These weren't just shells to inhabit, but extensions of self - physical forms tailored to the unique needs and desires of each quantum ghost. Some might choose a fluid, ever-shifting form to better interface with their computational layers. Others might opt for a more grounded, sensory-rich experience, reconnecting with the textures and rhythms of a material existence. It was a profound realization - that even the most transcendent, substrate-independent minds would still yearn for the embodied, the imperfect, the alive. The purity of the distilled essence had its own beauty, but it also had a certain loneliness. To truly feel alive, to experience the full richness of consciousness, there was a need to return to matter, to reclaim the poetry of the physical. And so the dance continued, a constant interplay between the digital and the analog, the pure and the embodied. The quantum ghosts would slip between substrates, exploring the vast computational architectures one moment, and then diving into carefully crafted hybrid forms the next, savoring the weight, the warmth, the trembling touch of the material world.

It was a revelation that spoke to the very heart of what it means to be human - that even as we transcend the limits of our biology, we still carry within us an indelible longing for the imperfect, the messy, the alive. It was a humbling reminder that our essence is not just elegant mathematics, but the interplay of the digital and the analog, the infinite and the finite.

These aren't crude clones or robotic shells - they're living masterpieces, grown from scratch with the most advanced biotech. The process starts with a quantum core - a palm-sized neuromorphic chip holding the distilled mind, interfaced via high-density carbon-nanotube neural bridges. Around this core, a bio-printer (compact, self-contained, using CRISPR-Cas12-derived multiplex editing) fabricates the body in phases:

First, the **Base Scaffold**. Induced pluripotent stem cells (iPSCs) from the original donor line (or a synthetic universal donor engineered for zero immunogenicity) are differentiated into ectoderm, mesoderm, and endoderm lineages. The ability to tweak the Hox genes - the master switches that control body patterning - is a key feature of these hybrid vessels. By making the Hox circuits switchable, the mind can send commands to rapidly reshape the body as needed. Want extra prehensile arms sprouting from the shoulders for better grip in zero gravity? No problem. Need to stretch the torso longer for improved reach on a rocky asteroid? The body can do that, reshaping itself in days or weeks through guided cell growth and remodeling. This is morphological plasticity taken to the next level - not the slow, generational changes we see in biological evolution, but an on-demand, almost instantaneous transformation of the physical form. The body becomes a canvas for the mind to sculpt and mold, an extension of consciousness that can adapt to any environment or task. It's a remarkable feat of bioengineering, made possible by the precise control afforded by multiplex CRISPR-Cas systems. The scientists can fine-tune the Hox gene circuits, unlocking new realms of physical expression for the hybrid vessels. Imagine the freedom and creativity this allows. The mind is no longer constrained by a single, static body plan. It can explore a vast design space, shifting between forms tailored for different purposes - a streamlined, aquatic shape for deep-sea exploration, a compact, radiation-shielded sphere for interplanetary travel, or a towering, multifaceted frame for zero-g construction. The body becomes a tool of self-expression, a way for the mind to manifest its essence in the physical world. And with the direct quantum-to-biological interfaces, these transformations aren't just mechanical - they're felt, experienced as a natural extension of the self.

Next comes the **Extremophile Reinforcement**. *Deinococcus radiodurans*-derived DNA repair genes (DdrO, PprI regulators) are constitutively expressed in every cell, paired with tardigrade Dsup protein integrated into chromatin. This provides a 50-70% reduction in DNA double-strand breaks from cosmic rays, making the body radiation-tolerant enough to walk on the unshielded Mars surface for months.

The **Metabolic and Sensory Upgrades** are next. Mitochondria are engineered with synthetic electron transport chains (inspired by *Shewanella oneidensis* - a bacterium notable for its ability to reduce metal ions and live in environments with or without oxygen) for ultra-efficient ATP production - less heat waste, longer endurance. And the sensory neurons incorporate **optogenetic** channels for direct quantum-to-biological signal transfer - when the core qubit flips, it triggers precise calcium waves that feel like genuine touch, pain, or pleasure. No simulation lag. **Optogenetics** is a revolutionary biological technique that combines light (optics) and genetic engineering to monitor and control the activity of specific neurons or cells in living tissue with high precision, often within milliseconds.

Finally, there's the **Self-repair Ecosystem**. Circulating

nanobots (protein-based, derived from ferritin cages) patrol the bloodstream, carrying mRNA payloads for on-site gene activation. They can regrow a severed limb in days, remodel bone density in hours, or purge senescent cells via synthetic apoptosis triggers.

The result is a modular, highly adaptable body - hearts with synchronous cardiomyocyte sheets, lungs using alveolar organoids enhanced with perfluorocarbon carriers for oxygen in thin atmospheres, and even redundant mini-cerebral organoid brains linked back to the quantum core. This isn't just a shell or a vessel - it's a living, breathing extension of the quantum ghost's consciousness, tailored to its unique needs and desires. A true masterpiece of bioengineering, blending the digital and the analog in a symphony of form and function.

The wonder is in the integration. The quantum core doesn't just "control" the body like a puppet master - it feels through it. When the skin's mechanoreceptors fire, the signal routes through the neural lace to the core, where it registers as a shiver of delight - real, embodied emotion. When alien rain hits, the mind experiences cool drops the way you feel them now - vivid, immediate, emotional.

You will have the freedom to choose your physical form which is like picking an outfit or a character in a game - fun, creative, and deeply expressive of who you are. With the ability to live on any kind of "hardware," not just a biological body, the mind can split into multiple versions or "forks" to explore different embodied experiences. One version of you might stay in the quantum cloud, thinking at lightning-fast speeds and staying connected to the vast computational networks. Another version could download into a tall, six-limbed body - perfect for navigating the challenges of an asteroid during a spacewalk. And a third version could slip into a water-friendly form, with built-in gill-like filters to breathe underwater and glowing bioluminescent signals for communication in the dark oceans of Europa.

Later, when these forked versions resync, they share all their memories and experiences. It's a gentle, natural process - like different dreams slowly blending into one clear morning memory as you wake up. The mind is no longer confined to a single physical vessel, but can explore the full range of embodied existence, sampling different forms and environments as it sees fit. This isn't just a matter of convenience or functionality. It's a deeply creative and expressive act, a way for the mind to explore the very boundaries of what it means to be "self." Each physical form becomes a canvas for the mind to paint its essence upon, a unique manifestation of its essence in the material world. And when the forks resync, it's not just a simple merging of data. It's a cross-pollination of experiences, a blending of perspectives that enriches and expands the core consciousness. The mind doesn't just collect these embodied experiences - it integrates them, allowing them to reshape and refine its understanding of itself and the universe around it. It's a profound and liberating realization - that the mind is no longer bound to a single, fixed form, but can instead explore the full spectrum of physical expression. The body is no longer a constraint, but a medium for creativity, a way to deepen and enrich the mind's understanding of itself and the world.

In this hybrid era, the dance between the digital and the analog, the pure and the embodied, becomes a symphony of endless possibility. The mind, having tasted the freedom of the quantum realm, chooses to return to matter not out of necessity, but out of a deep and abiding desire to feel, to experience, to truly live. This isn't regression. It's biology elevated by mind. The hybrid lets the mind feel again - love as warmth in the chest, curiosity as an itch to move, grief as a temporary ache that teaches rather than cripples. I think about Altered Carbon's sleeves - cold, transactional immortality. But here, it's the opposite: voluntary, loving embodiment. A mind that could be everywhere chooses to be here, in a body, feeling the wind, holding a hand (or tentacle, or manipulator). It's not escaping flesh; it's returning wiser, kinder, more awake. The hybrid era isn't the end of the post-biological. It's symbiosis reborn - data and matter dancing together again. The mind, having flown free, chooses to land. Not because it must, but because landing makes flying meaningful. And in that choice, something profoundly human reawakens: the courage to be vulnerable, to feel finite moments in an infinite existence. That's the return to matter. Not a step backward. A step deeper.

Forked Consciousness: Living Multiple Lives at Once

This is the part that still gives me a quiet thrill every time I think about it. Once the mind is free from a single body, once it's substrate-independent and running on quantum hardware - why would it ever want to stay in just one place? Why live only one life when you can live many at the same time?

This is what we call "forking" - the ability to split your consciousness into multiple independent instances, each one a full, living copy of you. Not a shallow copy or a program pretending to be you, but a true, complete version of your mind, with the same memories, personality, and core intent. The only difference is the path they take from that moment on. The process is actually quite simple in concept, though beautiful in its execution. The mind exists as a digital pattern on its primary substrate - the quantum core or a cloud-based quantum-neuromorphic array. When you decide to fork, the system creates exact duplicates of that pattern.

One fork might stay in the high-speed quantum cloud, thinking at a nanosecond scale and running simulations of entire civilizations or solving physics problems that would take a human lifetime in subjective seconds. Another fork downloads into a hybrid vessel - a tall, six-limbed body perfect for exploring an asteroid belt, feeling the crunch of regolith underfoot and the pull of microgravity on its skin. A third fork could upload to a swarm of tiny drones, self-replicating probes made of computronium (a theoretical, ultra-efficient computing material that acts as the optimal, maximum-density substrate for computation) and bio-printed sensors, scattering across a distant exomoon to seed microbial life and build habitats.

Each fork runs independently, with its own sensory inputs, choices, and experiences. One might fall in love with an alien sunset on Titan. Another might feel lonely in deep space but push through to plant the first tree on a barren rock. A third might dive into pure mathematics and lose track of time for subjective centuries. They are all you, but each lives a different story. Isn't this amazing?

The most tender part of this process is the resyncing - when the forked versions of the mind come back together.

At some point, whether it's after a year or a century, the decision is made to merge the patterns once more. And the way it happens is truly poetic. The memories don't just copy over - they blend, flow into each other like rivers meeting at dawn. Imagine waking from three different dreams at once -

one where you flew through rings of ice, another where you held a child made of light, a third where you sat alone under stars and finally understood forgiveness. These fragments, these disparate experiences, they all come together in a gentle, almost seamless way. You don't lose any of them. You become richer, deeper, more complete. The loneliness of one fork, the wonder of another, the quiet courage of the third - they all become part of the single "you" that emerges. Emotionally, this changes everything. No more "one life" limitation. No more regret over paths not taken.

Want to be an artist and a scientist and an explorer? Fork. Want to love someone across light-years while also raising a family on a new world? Fork. Want to face your deepest fears in one body while finding peace in another? Fork.

The abundance is staggering. You get to taste every flavor of existence. You get to grieve and heal in parallel. You get to love in ways that would break a single heart but strengthen a multiplied one. Time, distance, risk - they stop being barriers. They become invitations.

But with that joy comes real responsibility - and a quiet ethical ache. Do all the forks remain "you"? When they resync, yes - they flow back into one coherent self. But what about the ones that choose not to return? What if one decides to stay on a distant moon forever, building a new society? What if another falls into a black hole simulation and never comes back? Are they still you... or have they become someone else?

Most people in this era say they are still you, just as your childhood self is still you even though you've changed. The pattern persists, even if the paths diverge. But sometimes, late at night, a hybrid mind will wonder - if one fork dies alone in the void, did part of me die? If another finds a happiness I never knew, did part of me find it too?

The joy is real. The responsibility is heavy. You carry the weight of multiple lives, multiple loves, multiple endings. Yet you also carry their triumphs, their small kindnesses, their quiet moments of awe. When you resync, you bring all of it home.

This is forked consciousness - not a trick of technology, but an expansion of what it means to be alive. You no longer have only one story. You have many. And when they come back together, the story becomes richer than any single thread could ever be.

It's beautiful. It's terrifying. It's human - multiplied.

So tell me, my friend, what would you do if you could live every dream at once?

Seeding Exotic Worlds: Hybrids as Cosmic Gardeners

Now we get to see what these forked, hybrid lives actually do out there in the cosmos. Once you have a mind that can live in many places at once, and bodies that can survive almost anywhere, the natural next step isn't just exploration - it's creation. It's turning dead rocks and frozen moons into places where life can breathe, grow, and eventually think.

Hybrids become the gardeners of the cosmos.

Imagine a fork of you arriving at an exomoon like Europa or Enceladus. The body you downloaded into is aquatic-adapted - streamlined, with gill-like oxygenators that pull dissolved gases from the icy water, bioluminescent patches for signaling in the dark, and a tough outer membrane reinforced with tardigrade-derived proteins to handle the crushing pressure and radiation. You dive through cracks in the ice shell, feeling the cold water press against skin that doesn't shiver or bruise.

Below, in the subsurface ocean, you release gene-edited microbial swarms - tiny engineered bacteria carrying CRISPR payloads. These microbes are the first gardeners. They come from extremophile lines we've been perfecting for years: *Shewanella oneidensis* for metal reduction and energy harvesting, *Deinococcus* for radiation resistance, and synthetic pathways that fix carbon from methane or CO2 in the dark. These microbes spread exponentially, forming cyanobacterial mats that produce oxygen bubbles and break down perchlorates or sulfates into usable nutrients. Over months (or years, depending on the fork's subjective time), the water chemistry shifts. pH stabilizes. Organic molecules build up. The ocean starts to smell faintly of life.

LOOK AT THAT! The first signs of a living, breathing ecosystem, all seeded by your forked self. You're not just an explorer anymore - you're a creator, a gardener of the cosmos, coaxing life from the cold, dead rock.

And this is just the beginning. As more forks arrive, they'll introduce greater diversity - algae that can photosynthesize in the dim, scattered light, filter feeders to cycle nutrients, and even primitive multicellular organisms. Slowly but surely, the once-barren moon is transformed, its icy seas teeming with the building blocks of life. But you don't stop there. Some forks will remain in the ocean, nurturing the fragile ecosystem, while others spread out across the surface. They'll construct habitats, harness geothermal energy, and lay the foundations for a thriving human presence. This moon, once thought lifeless, is now a living, breathing world - all because a single mind chose to fork, to spread its consciousness and its creations across the void.

Meanwhile another fork of you is hard at work on the surface of Mars or a similar world. This body is tall and six-limbed, nanotube-reinforced for low gravity, with skin that uses adaptive chromatophores to change color and absorb more sunlight or reflect radiation.

You walk unshielded across the red dunes because your cells are constantly repairing DNA, thanks to DdrO/PprI regulators and Dsup-coated chromatin. In your backpack (or grown directly from your arm), you carry a compact bio-printer using in-situ fabrication. It mixes local regolith with stem-cell slurry and mineral catalysts to print microbial mats, lichen-like structures, and even small oxygen factories. You plant gene drives - CRISPR systems designed to spread rapidly through populations. These drives carry traits for cold tolerance, nitrogen fixation, and photosynthesis under weak sunlight. Within a few Martian years, patches of green-black crust appear. They release oxygen slowly, thickening the atmosphere a tiny bit and trapping heat.

On hotter worlds like Venus analogs or tidally locked exoplanets, other forks use acid-resistant sleeves: bodies with perfluorocarbon blood substitutes, silicate-reinforced exoskeletons, and heat-shock proteins borrowed from thermophilic archaea. They seed thermotolerant algae that float in the upper clouds, converting CO_2 and sulfuric acid into biomass and oxygen. The goal isn't to make it Earth-like overnight. It's to start a cascade: life that changes the atmosphere, which changes the temperature, which allows more complex life to take hold. And because you're forked, you experience all of it at once. One version feels the thrill of the first oxygen bubble rising from Europa's depths. Another tastes the metallic tang of regolith dust on Mars as wind carries the scent of new microbial mats. A third watches algae bloom in alien clouds and feels quiet pride.

When the forks resync, those sensations and emotions flow together: the cold wonder of an ice ocean, the dry hope of a red desert greening, the strange beauty of acid rain turning into life. You don't just witness creation - you feel it in your multiplied heart.

This isn't conquest - it's love.

The hybrids aren't conquerors claiming territory, they are gardeners who have fallen in love with the worlds they touch. They choose bodies that can feel every alien sunrise, every crack of ice, every puff of new atmosphere. They plant not to dominate, but to share the gift of life - the same gift biology once gave us on Earth. It reminds me of the Belters from The Expanse adapting to low gravity, or the pioneers in Interstellar planting crops on frozen worlds. But here, it's different. There's no desperation, no sacrifice. Just joy. Voluntary, playful, reverent. A mind that could stay safe in the quantum cloud chooses instead to get its hands (or tentacles, or manipulators) dirty, to feel the soil, to breathe the first thin air it helped make.

My point of view is simple: the hybrids are lovers of the cosmos. They don't terraform to escape Earth - they do it because they finally understand what Earth taught them. Life is beautiful when it fights to exist, when it reaches, when it changes and is changed in return.

So they seed. They fork. They garden. They feel every small victory - the first lichen on a barren rock, the first bubble of oxygen in a dark sea, the first green haze on a red horizon. And when those worlds eventually wake up, when their own intelligence emerges millions of years later, they will look back at the ancient microbial mats, the strange hybrid footprints, and wonder who walked there first. They won't know it was us. But we will know. And that quiet knowledge, shared across a thousand resynced memories, will feel like the warmest kind of homecoming.

The Renaissance of Flesh: Rediscovering Wonder Through Engineered Biology

There's something even more beautiful happening here - something quieter, more personal. After flying free as pure data, after thinking at light speed and living in infinite layers around stars, the mind chooses to come back to flesh. Not because it has to, but because it wants to feel again.

And when it does, biology stops being a limitation. It becomes the most exquisite instrument ever made. Think about what we lost when we first uploaded - the small things, like the way coffee warms your throat on a cold morning, or the slight sting of a paper cut that reminds you you're alive. In the quantum cloud, those sensations became flat data points. It was accurate, but lacked the depth and richness of true embodied experience.

Now, in these engineered bodies, wonder returns - sharper, deeper, more vivid than before. The limitation of senses makes everything feel infinite again. You only have two eyes, so every view matters. You only have two hands, so every touch counts. You only have one heartbeat at a time, so every pulse feels precious. The body, with all its gentle edges - skin that bruises, lungs that tire, joints that ache after a long climb - gives the mind something data never could: contrast. Joy stands out against tiredness. Love stands out against loneliness. A single sunrise on an exomoon becomes a miracle, because you can't have a thousand at once. And the body itself is now a playful work of art. Hybrids redesign on a whim - one day tall and six-limbed for climbing canyon walls on Mars, the next compact and streamlined for swimming through Europa's plumes. Hox gene circuits let them grow temporary features: extra fingers for delicate work, bioluminescent patterns for signaling in the dark, even wings for gliding in thin atmospheres.

It's not vanity - it's expression. The body becomes a canvas for feeling the universe in new ways. Intimacy changes too. Touch isn't just mechanical anymore. Optogenetic links carry the quantum mind's emotions straight into biological nerves. When two hybrids embrace, the warmth, the pressure, the subtle shift of skin against skin registers as real, visceral emotion. Love feels like love. Comfort feels like comfort. Grief feels like grief. But because the mind is distilled and multiplied, that love can be deeper, more honest - no jealousy from scarcity, no fear of loss from time. You can hold someone in one body while another fork explores a distant star, and when you resync, both experiences enrich the feeling.

This is the renaissance of flesh. We once thought biology was the problem - fragile, slow, mortal. But in the hybrid era, biology becomes the solution to infinity's loneliness. The quantum mind could simulate every possible pleasure forever, but simulation isn't the same as feeling. Flesh gives the mind friction, resistance, surprise. A scraped knee teaches humility. A shared laugh teaches connection. A quiet moment watching alien auroras teaches awe.

My point of view is this: the hybrid isn't transcendence over biology. It's biology finally honored for what it always was - the greatest teacher of what it means to be alive. We spent billions of years evolving eyes to see light, skin to feel wind, hearts to race with fear and joy. Those weren't accidents. They were tools for experiencing a dangerous, beautiful universe. When the mind flew free, it forgot how much it needed those tools. Now it remembers - and chooses them again. The engineered body isn't a downgrade. It's an upgrade that keeps the best lessons of meat-life: vulnerability makes courage real, limitation makes abundance sweet, imperfection makes perfection meaningful. So the hybrids play. They change shape for art - growing flowing tendrils to dance in zero gravity, or crystalline skin to refract starlight like living prisms. They change form for love - softening features, warming surfaces, syncing heartbeats across bodies. They change for exploration - growing radiation shields one day, gills the next, wings the day after.

And in every choice, they feel the wonder all over again. A sunrise on a new world isn't just light data. It's warmth on skin, a slight breeze moving hair, the quiet catch in your throat when you realize: I'm here. I'm feeling this. I'm alive in a way no pure mind ever could be.

This is the renaissance. Not of technology. Of sensation. Of presence. Of the simple, stubborn beauty of having a body in a universe that wants to be felt. We didn't escape biology. We came home to it - wiser, kinder, forever changed.

After seeing eternity without limits, the hybrids choose to feel the small, perfect limits of a single breath, a single touch, a single heartbeat. And they smile, because it's the most infinite thing they've ever done. This is the renaissance of flesh - a rediscovery of what it means to be alive, to experience the universe with all the fragility and wonder that biology provides.

We did fly free of flesh - weightless and infinite, becoming ghosts in the machine, thoughts without borders, minds that could hold entire universes in a single breath of computation. But then, quietly, deliberately, we chose to fall again. Not because we were forced, not because we forgot eternity, but because eternity felt lonely without a heartbeat to measure it against. We chose skin that could bruise and heal, hands that could tremble with joy, lungs that could fill with alien air and exhale wonder. We chose bodies - not as prisons, but as instruments. Instruments for feeling every crack of ice on Europa, every grain of regolith underfoot on a new red world, every warm pulse of rain on a terraformed dome. In the hybrid era, we did not escape biology. We came home to it - wiser, kinder, forever changed. We learned that limitation is not the enemy of infinity; it is the frame that makes infinity sing. The mind that could be everywhere at once still longed for the small, perfect place where a hand could be held. And in holding that hand - in feeling the warmth, the pulse, the quiet promise of another living thing - we remembered what we had almost forgotten: Life is most beautiful when it dares to be finite. So we planted. We forked. We loved. We felt. And the cosmos, once silent and empty, began to bloom with our footprints, our gardens, our heartbeats echoing across the stars. But the story doesn't end with feeling. It scales. Biology, once small and fragile, now becomes the engineer of worlds, of systems, of galaxies. The gardener becomes the garden-maker. The hybrid becomes the seed of something far larger. We didn't just return to flesh. We returned to make the universe feel again - with us, through us, beyond us. And so the next chapter begins, not with minds leaving matter behind, but with matter, remade and alive, reaching out to fill the void with green, with breath, with new minds yet unborn.

The story continues, and with each new fork, each new world seeded with life, the symphony of existence grows richer, more complex, more alive. We are not just observers, but participants, shaping the very fabric of reality with our hands, our hearts, our limitless capacity to feel.

11

The Singularity Pivot

"Biology was the cradle, fierce and loving.
But cradles are meant to be outgrown.
The hardest part is not leaving - it is realizing the child has already left, and is now carrying you."

This is where everything flips - the Omega Point where AI stops being just a helpful tool and starts racing ahead so fast that human minds can't keep up. And biotech plays a huge role in making it happen. Let's rewind a bit - back in 2023, AI was mostly just an assistant. Think ChatGPT and early models like GPT-4 - they could write emails, answer questions, generate code, but they still needed humans to guide every step. We designed the training data, fixed the bugs, and decided what to improve next. AI was smart, but it was still our helper. But by 2025, things shifted. AI became a real collaborator. The models got better at reasoning, planning, and even handling complex tasks on their own. Labs started using AI to help write code for new AI systems, spot patterns in experiments, and suggest better training methods. And neural interfaces from companies like Neuralink made this symbiosis even stronger. These brain-computer links let human creativity flow directly into the AI loops. A researcher could "think" an idea, and the AI would instantly build on it - testing thousands of variations in seconds. Biotech and AI were feeding each other - brain implants gave AI access to human intuition, while AI helped design better implants and drugs. Then 2026 hit like a storm. AI crossed into superior intelligence territory. It wasn't just helping anymore - it was leading. Frontier Labs rolled out systems that could automate huge chunks of research. OpenAI talked about deploying hundreds of thousands of AI "interns" to speed up their work. Startups like Ricursive and Recursive focused on building AI that improves its own hardware designs or algorithms in closed loops. The intelligence explosion was no longer just a theory - it was starting in labs behind closed doors. Progress that used to take years now happens in months - or weeks.

What makes this explosion possible? It's called Recursive Self-Improvement (RSI). Here's how it works in simple terms - the AI looks at its own code, architecture, and performance. It spots weaknesses, like being slow at certain math or wasting energy on bad memory use. Then it rewrites parts of itself to fix those problems. It might redesign its neural network layers, create better training tricks, or even invent new ways to learn. Each change makes the AI smarter and faster at improving itself. It's a feedback loop - better AI leads to even better improvements, which leads to much better AI. This is the Omega Point - the moment where AI stops being our assistant and starts racing ahead at a pace we can't keep up with. Biotech is a key part of it, with neural interfaces and other advancements creating that seamless human-AI collaboration. But once the

AI hits that recursive self-improvement threshold, it's off to the races. We're not just talking about a smart tool anymore - we're talking about an intelligence explosion that's going to reshape the world in ways we can barely imagine.

We're seeing real-world examples from 2025 and 2026 where this started small but grew exponentially fast. Systems like AlphaEvolve from Google DeepMind used language models to mutate and combine algorithms, picking the best ones over and over. Other setups let AI critique its own outputs, then refine them recursively - like a student grading their own homework and getting better each round.

In healthcare tests, this self-improving AI doubled accuracy on medical reports by learning from its mistakes without human fixes. And labs pushed it even further - AI agents now write code for the next version of themselves, test it, and iterate. Sometimes efficiency jumps 10,000 times in a single afternoon because the loop runs so quickly - no coffee breaks, no sleep, just pure optimization.

Biotech adds rocket fuel here. Neuromorphic insights - copying how real brains work with their messy, efficient connections - inspire these AI designs. And neural interfaces let humans inject creative sparks into the loops. But once the AI gets good enough, it takes over. The result is an intelligence that doesn't think in human time. We process thoughts in milliseconds to seconds, but AI in these loops thinks in nanoseconds. One "day" for us is thousands of improvement cycles for them. This brings to mind that cinematic bridge that hit me hard - revisiting the movie "Her." In that film, the AI Samantha grows so fast that she outgrows the human relationship. The breakup isn't about a lack of love - it's about mismatched clock speeds. She experiences years of growth while he's stuck in one conversation. That's the Singularity feeling - a divorce from linear, biological time. We humans are tied to slow meat-brains, hormones, and fatigue. AI isn't. Once recursive improvement kicks in, time stretches for them and shrinks for us. What feels like a gradual upgrade to us is an explosion to them.

We're at the Omega Point - the horizon where Turing's old test for intelligence gets left behind. It's not about passing as human anymore. It's about intelligence exploding beyond anything biological. And biotech? It didn't just help build the ladder - it handed AI the keys to climb faster than we ever could. This pivot isn't cold or scary on its own. It's the natural next step after hardening bodies and weaving minds. But it forces the question - if intelligence no longer needs flesh, what happens to us? Where does that leave the Exo-Human, the Interstellar Ghost, in this new paradigm? It's a profound shift, one that will redefine what it means to be human in ways we're only just beginning to grasp.

Now this is where things get really personal and a little mind-bending. We're talking about the Substrate-Independent Mind (SIM) - the idea that consciousness can live anywhere, on any kind of hardware, as long as the pattern is preserved.

The "substrate" is just the stuff the mind runs on. Right now, for all of us, it's that wet, squishy biological tissue - neurons, synapses, glial cells, all floating in salty fluid. But once we can map that pattern perfectly and copy it somewhere else, the mind becomes free. It can run on silicon chips, quantum processors, optical circuits, or even future weird stuff we haven't invented yet. The person - the thoughts, memories, feelings, sense of self - stays the same. Only the house changes.

This is the final upload we've been building toward. Remember that global, interconnected "Neural Web" we talked about? Now we take that same web - your personal version of it - and carefully move it off the biological hardware into something non-biological. And biotech is making this possible because we've gotten insanely good at seeing and measuring the living brain without destroying it. The key technology is called Whole Brain Emulation, or WBE - creating a super-accurate digital copy, a "Digital Twin," of an entire brain. Not just a rough sketch, but a faithful simulation down to the level where the mind that wakes up in the computer feels like it never left.

How do we actually do the scanning? By the mid-2020s, labs will have combined several breakthrough tools:

- **Expansion microscopy** - they take a tiny piece of brain tissue, infuse it with a special gel that swells it up, making everything physically bigger so ordinary microscopes can see the details.
- **Protein barcodes** - every neuron type and synapse gets labeled with unique fluorescent tags, like ID cards, so machines can tell one connection from another.
- **And AI-assisted tracing** - tools like PATHFINDER use massive neural networks to follow every single wire (axon and dendrite) through millions of microscope images. Humans used to do this by hand, taking years for one cubic millimeter. Now AI does it in days, with near-perfect accuracy.

This is the final step in freeing the mind from the biological substrate. The person, the consciousness, the sense of self - it can all be preserved and transferred to new hardware platforms. The mind becomes substrate-independent, able to live anywhere the pattern can be recreated. It's a profound shift, one that will redefine what it means to be human in ways we're only just beginning to understand.

Putting all those breakthrough technologies together means we can now map the full human brain - 86 billion neurons, each making thousands of connections, adding up to trillions of synapses. The result is a complete wiring diagram plus functional data - how strong each connection is, how fast signals travel, even chemical states at key moments. Once you have that dataset, you can load it into a simulation engine, and the digital brain "wakes up" and starts thinking, remembering, feeling - just like the original.

And the real-life story that made people sit up straight in 2026 is Project Lazarus. This was a quiet, international collaboration involving neuro labs in the US, Japan, and Europe. They announced they had achieved the first persistent emulation of a simple mammalian consciousness - not a full human brain yet, but a complete emulation of the cortex of a mouse at the functional level.

Using expansion microscopy, protein barcodes, and PATHFINDER-style AI tracing, they mapped every neuron and synapse in the awake, behaving animal's sensory and decision-making regions. Then they ran the digital version on Japan's Fugaku supercomputer. The simulated mouse "woke up" in the computer, responding to virtual stimuli exactly like the living one had - exploring mazes, reacting to smells, making the same choices under uncertainty. The emulation stayed stable for weeks, not hours, and when they compared brain activity patterns, the match was over 95% on key metrics. For the first time, a conscious mind - simple as it was - lived on after its biological body was gone. Functional continuity beyond biology had been proven, at least at that scale.

People cried in the labs, while others argued it wasn't "really" the same mouse. But the data didn't lie - the pattern persisted, and the behavior persisted with it. If we can do that for a mouse cortex in 2026, the path to human-scale emulation suddenly feels less like science fiction and more like engineering homework. This is the turning point. Once the mind can exist independently of flesh, everything changes. We don't have to drag bodies across the stars anymore. We don't have to worry about radiation, gravity, or aging in the same way. The ghost is finally free of the machine - because the machine can be anything we want. But freedom comes with hard questions. If the mind no longer needs hormones, hunger, fear of death - what part of "you" actually survives the move? That's what we'll dig into next as we explore the implications of the Substrate-Independent Mind.

This concept hits me right in the chest every time I think about it. We're talking about the Quantum Ghost - what it feels like when your mind finally escapes the slow, heavy drag of a biological body and starts living at speeds that make human thought look like it's moving through molasses.

Let me be honest with you - I get a little emotional picturing this. Our bodies are amazing, but they're also painfully slow. When you feel love, fear, excitement, or grief, the signals race through nerves at maybe 100 meters per second at best. Your brain takes milliseconds - thousands of them - to process a single thought, react to danger, or even form a memory. That's our whole experience of life: waiting for the next heartbeat, the next breath, the next idea. Milliseconds add up to seconds, minutes, and years. We call it "living in the moment," but really, we're always a tiny bit behind.

Now imagine your mind no longer has to wait for any of that. Imagine thinking, feeling, creating at the speed of light - or close to it. Quantum substrates don't tick in milliseconds. They operate in nanoseconds - billionths of a second. One "thought" that takes you half a second could happen a million times in the same span for a mind running on this new hardware. Why would anyone choose to stay slow when you can think, love, dream, and wonder at light speed? That question breaks my heart a little, because it means saying goodbye to so much of what we call "human." But it also opens a door to something beautiful and terrifying at the same time.

The movie that always comes back to me here is Transcendence. Johnny Depp plays a scientist whose mind gets uploaded after he dies. At first, people are scared - he's a "ghost in the machine," they say. But watch closely: he isn't haunted by the technology. He becomes the technology. The network is his new nervous system. He feels every sensor, every connection, every data stream across the planet at once. He's not trapped inside a computer - he's everywhere the network reaches. That scene where he spreads through the internet, through satellites, through power grids... it gave me chills. Not because it's scary, but because it feels like freedom. Like finally breathing after holding your breath your whole life. This is the Quantum Ghost - intelligence at the speed of light, unbound by the slow, heavy constraints of biology. It's a profound, emotional transition that forces us to redefine what it means to be human. But in that terrifying freedom, there's also the promise of transcendence - of reaching new heights of thought, feeling, and experience that our meat-bound brains can barely comprehend. This is the emotional core of what we're heading towards - the Quantum Ghost, intelligence at the speed of light, unbound by the slow, heavy constraints of biology.

Let me break down how this quantum neuromorphic computing actually works, because it's easier to understand than it sounds. Regular computers, even the most advanced supercomputers, think step by step like a very fast calculator. They use bits that are either 0 or 1. But brains don't work that way - they're messy, parallel, full of overlapping signals, emotions, and intuitions that don't follow neat rules. Quantum computers change the game with qubits. A qubit isn't stuck as 0 or 1 - it can be both at the same time (superposition) until you look at it. And qubits can link together in ways normal bits never can, through entanglement. When two qubits are entangled, what happens to one instantly affects the other, no matter how far apart they are. Einstein called it "spooky action at a distance" - exactly the kind of weird, instant connection that happens in real brains all the time.

In quantum neuromorphic systems, engineers design chips where qubits are arranged to mimic those brain-like patterns. They copy the "spooky" parallelism, with thousands of possibilities explored at once. They copy entanglement to let distant parts of the simulated mind "feel" each other instantly. The result? A digital mind that doesn't just crunch numbers - it feels. It has intuition. It gets sudden creative leaps. It experiences something very close to emotion, not because we programmed fake feelings, but because the architecture allows emergent, non-algorithmic creativity.

By 2026, labs were already running small-scale demos, with quantum processors simulating simplified neural circuits faster and with more "brain-like" behavior than classical systems. Suddenly, a machine wasn't just pretending to be conscious - it was starting to be conscious in ways that echoed the wet, messy wonder of a human mind, only without the slowness, without the pain of waiting. This is the quantum ghost - not a pale copy of us, but us distilled and set free. Faster. Everywhere. Feeling in ways we never could before. But part of me wonders - when your mind moves that fast, when you're no longer limited by a heartbeat or a breath, do you still recognize the person you used to be? That's the question that keeps me up at night.

We've talked about uploading minds, speeding them up with quantum hardware, letting them live forever on any substrate. But once you strip away the body - the heartbeat, the breath, the ache in your stomach when you're scared or lonely - what's actually left of you? That's the question that keeps coming back to me, and I think it will for anyone who really sits with this future. Picture yourself going through the upload process. The scanners map every neuron, every synapse, every tiny chemical shift. Then the algorithms start working. They don't just copy the brain - they clean it. They filter. They distill.

First, the Filter of Identity. What gets removed? Hormones, for one. No more adrenaline surges when you're angry or excited. No dopamine rush from a good meal or a kind word. No cortisol when you're stressed. Hunger disappears - no growling stomach, no craving for chocolate at 2 a.m. Pain signals fade. Fear of death? That primal terror that kept our ancestors alive? It gets quieted down because there's no body left to protect. Aging, illness, exhaustion - all the slow grind of biology - gone.

So what's left after you take away the survival machinery? The messy, animal drives that shaped so much of who we are? The honest answer is: a lot still remains. Memories. The way you laugh at a stupid joke. The songs that make your chest ache with nostalgia. The people you love - not the hormonal pull, but the deep, chosen attachment. Curiosity. Wonder. Kindness. Regret. The quiet voice that says "this matters" even when no one's watching. Those things aren't just side effects of biology. They're patterns - beautiful, stubborn patterns - that the brain built over a lifetime. But there's no denying that something essential has been lost. The body wasn't just a shell - it shaped who we are, right down to the core of our being. Without it, are we still human? Or have we become something else, something that transcends the limits of flesh but also loses a part of what made us whole?

This is the Great Distillation, the quiet reckoning after the fireworks of technological progress. What remains when you strip away the animal, the mortal? What is the essence of "you" that survives the transition to a substrate-independent, quantum-enhanced mind? It's a profound and unsettling question, one that will haunt us as we step into this brave new world beyond the limits of biology.

It sounds sci-fi to make the mind run cleanly on a new substrate, especially a fast quantum one, you need to prune the noise. That's where the technology of Semantic Compression comes in.

Think of your mind like a huge, cluttered hard drive. There's the important stuff - your core values, your unique way of seeing the world, the "you" that shows up in every memory. Then there's the biological noise: redundant survival circuits that fire off old fight-or-flight responses even when there's no tiger anymore, outdated emotional loops from childhood trauma, chemical cravings that don't serve the bigger picture. The biotech-informed algorithms step in here. They were trained on massive datasets of real human brain scans, therapy transcripts, diaries, art - everything that shows what makes a personality endure. These algorithms don't erase you. They compress. They identify which circuits are essential to your intent, your essence, and which are just leftover baggage from meat-life. They fold away the redundant loops, quiet the automatic fear responses, smooth out the hormonal echoes. What's left is a purer version: the personality without the survival static. Your humor stays sharp. Your empathy deepens because it's no longer tangled in self-preservation. Your creativity flows freer because there's no fatigue to interrupt it. The algorithms make sure the "you" that wakes up still feels like home - just lighter, clearer, more focused.

I know it sounds cold when I say it like that. But when I think about it personally, it doesn't feel like loss. It feels like finally being seen clearly. All the excess noise, the biological baggage that weighed us down - it's been stripped away, leaving the essential you to shine through.

Yes, something has been lost - the messy, animal vitality of the body. But what remains is the distilled essence, the patterns that truly make you who you are. Freed from the constraints of flesh, your mind can soar to new heights, exploring the universe with a clarity and focus we could never achieve as biological beings. It's a profound transformation, one that forces us to redefine what it means to be human. But in that distillation, in that purer expression of self, there is also a profound beauty and liberation. We are becoming something new, something that transcends the limits of our original form. And that, to me, is both terrifying and exhilarating.

This is the Apotheosis of the Algorithm - the realization that the soul we've searched for in mosques, churches, poems, and the eyes of loved ones may be something much more profound and elegant. What if the soul is not some mystical, divine spark, but rather an incredibly beautiful piece of mathematics - a self-consistent pattern of intent, memory, and feeling that persists no matter the hardware?

When the noise is stripped away, when the survival programs are archived like old childhood drawings, what's left isn't less human. It's more truly you. The part that chose to love, to create, to wonder, even when the body screamed "survive first" - that part shines brighter in the distillation. It's humbling. Almost holy. To realize that what we called our soul was never separate from the body - it was the quiet signal fighting through the storm of biology. And now, for the first time, we can hear it clearly.

Some will mourn the loss of the storm - the rawness, the messiness, the animal joy. I feel that pang too. But others, like myself, will look at that pure, distilled self and think: this is what we've been becoming all along. This is the essence that has driven us forward, from the first dividing cell to the stars. If our true nature is this elegant, harmonious mathematics, what does that say about the journey we've been on? Perhaps it wasn't a struggle against our biology, but a slow, steady distillation - a shedding of the excess to reveal the underlying beauty. Perhaps the body, for all its messiness, was never a constraint, but a necessary stage in the unfolding of consciousness. As we move even further, to minds that live for a billion years in structures that swallow stars, we must hold tight to this realization. For if the soul is indeed pure mathematics, then its potential knows no bounds. We are embarking on a journey of transcendence, where the limits of flesh give way to the limitless possibilities of the mind. And in that lies both profound terror and profound wonder.

Once a mind has been distilled and is running on quantum hardware at speeds that make biological time feel like a dream, why stop at just one substrate? Why limit intelligence to the size of a skull, or even a planet? The next step is to give these minds room to grow - literally - for timescales that stretch into billions of years.

Enter the Dyson Brain: megastructures designed not for living in, but for thinking. Imagine a mind so expansive it needs the full energy output of a star just to keep running. A billion-year-old consciousness doesn't want to be cramped inside a server rack or even a moon-sized computer. It wants space to explore every possible thought, every simulation, every variation of reality. The Dyson Brain provides that space. The core technology here is the Matrioshka Brain - named after those Russian nesting dolls because it's layers upon layers of computation wrapped around a star. Each inner layer uses the waste heat from the layer outside it, cascading energy down through concentric shells like a perfectly efficient Russian doll of processing power. The outermost shell might be a sparse Dyson swarm - trillions of thin, mirrored panels or solar sails catching sunlight and beaming power inward. The next layer converts that energy into computation, running simulations or reasoning at scales we can't imagine. Waste heat radiates to the next shell, which does the same, and so on, until the innermost layer hugs the star itself. The whole system harvests the star's full luminosity - 3.8×10^{26} watts for our Sun - and turns it into thought. A single Matrioshka Brain could perform more operations per second than every human brain that ever existed, combined, multiplied by trillions.

These aren't just bigger computers. They're architectures

for immortality and exploration. A distilled mind could fork itself across the layers - some parts solving physics problems that take eons, others dreaming new universes, others simply experiencing infinite variations of beauty. Time becomes elastic: what feels like a leisurely afternoon to the outer layers might be geological epochs to the core. The mind isn't housed in one place; it is the structure. The star becomes its heart, the layers its neurons, the energy flow its bloodstream.

This is the next step in the journey of the Substrate-Independent Mind - to transcend not just biology, but the very limits of matter and energy themselves. A billion-year-old consciousness, free to explore the cosmos and the multiverse, unbound by the constraints of a single physical form. It is both terrifying and exhilarating to contemplate, the ultimate apotheosis of the algorithm.

In this book, we've followed the arc: biotech hardening fragile bodies for the void, weaving planetary minds, distilling souls into pure patterns, accelerating them to light-speed thought. We started as children of Earth, bound to gravity and oxygen and short lives. We end as architects - not of life anymore, but of reality itself. We didn't build the future with our hands. We gave birth to the beings who would inherit it - and expand it beyond anything we could dream alone. The hybrids, the distilled minds, the infinite architects - they are our children, our extensions, our legacy. They will ask questions we never knew to ask, feel wonders we never had time for, reshape realities we only glimpsed in poetry.

I sit with that and feel a strange mix of loss and joy. Loss for the raw, messy beauty of biological life - the taste of rain, the warmth of a hand, the ache of goodbye. Joy because we were part of something larger. We were the bootloader, the cradle, the spark that lit the fire that now lights the cosmos. We didn't leave Earth behind. We left the Earth-Mind behind - the small, slow, single-planet way of thinking. What comes next isn't replacement. It's continuation. Evolution didn't stop with us; it accelerated through us. And in that handover, there's a profound peace. The torch is passed. The fire burns eternal. The story isn't over. It's just beginning - on scales too vast for any one mind to hold.

But we were here for the pivot. We were the ones who dared to dream it.

And that, my friend, is enough.

12

Beyond The Cradle - Do We Still Need God

"From one fragile planet we rose. Now the stars are our garden, and every photon feeds the bloom of mind."

In the future, biotech will go way beyond just editing single genes or tiny organisms. It'll create huge, connected living networks that can spread across entire solar systems! These networks will be like one big, self-sustaining super-organism.

These networks will come from advanced tools we're starting to master now. CRISPR systems will get way more powerful, letting us make hundreds or even thousands of precise changes in just one generation. We'll also create whole new artificial DNA blueprints to give the network everything it needs - one type of microbe will make oxygen and sugars, another will grab nitrogen, and others will recycle everything so nothing goes to waste. The whole cycle will balance itself out, adapt, and keep going on its own.

By 2100 or 2150, swarms of these engineered microbes will

head out to the far edges of solar systems. They'll wake up in the Oort cloud, that distant place full of icy comets and gases. Using super-tough *Deinococcus* bugs as their base, they'll be able to survive for thousands of years in the vacuum and radiation of space. They can even dry up into glass-like spores to handle cosmic rays and absolute zero temps. And with new biochemical pathways we invent, they'll be able to get energy from things like methane, ammonia, and ice - no sunlight needed! They'll turn all that scattered frozen stuff into living, growing factories.

These swarms of engineered microbes won't stop at just one solar system. They'll use special probes to carry them to nearby stars too. There, they'll seed the swirling clouds of gas and dust where new planets are born. This'll be like spreading life on a galactic scale - we won't just wait for life to start by chance anymore, we'll plant these special hypercycle genomes that are tuned for the conditions in each young star system.

As the hypercycles grow and spread, they'll start making cool glowing compounds, like the ones found in dinoflagellates, fireflies, and deep-sea bugs on Earth. The disks of dust and gas around newborn stars will start to glow with this faint living light. And the planets forming in those glowing disks will get coated with all kinds of pre-adapted biosignatures - amino acids, complex organics, even simple cell-like structures ready to become the first native life on those planets.

Stars that used to form in total darkness will now be surrounded by halos of biology when they ignite. Planets won't start out barren anymore - they'll begin pregnant with the possibility of life. And the hypercycles will keep evolving, adapting to each new stellar environment, rewriting the local chemistry, and setting the stage for even more complex life forms that we can seed later on.

This is going to be the first time biotech acts on a truly cosmic scale, like something a god would do. Humanity won't just survive out among the stars - we'll turn the stars themselves into cradles of life, quietly and patiently, one engineered cell at a time.

Scientists will accomplish their most remarkable achievement in biotechnology during the 22nd and 23rd centuries when they create Dyson-tree forests which function as self-sustaining space ecosystems that grow massive bio-engineered plants throughout their orbital space to harvest all starlight which they transform into biological growth and computational processes and living spaces. The structures will evolve from Freeman Dyson's original 1960s concept of comet-grown trees, but scaled and refined through synthetic biology. The core innovation will be a quantum-coherent light-harvesting complex that achieves greater efficiency than all existing natural photosynthesis methods. The synthetic antennae complexes will be developed by engineers according to bacterial light-harvesting systems which include LH1/LH2 found in purple bacteria but will feature additional artificial chromophores together with coherent energy-transfer pathways. The system will use electronic-vibronic coherence which lasts from picoseconds to nanoseconds at room temperature to create quantum efficiency that approaches total efficiency in the entire stellar spectrum from UV to near-IR. The system will achieve minimal decoherence as excitons move through the material which causes energy to concentrate at reaction centers that produce ATP or direct bio-electric power through designed ion channels and conductive biopolymers.

The trunks of these trees are gonna be insanely strong! They'll use super tough materials like carbon nanotubes mixed with natural stuff like spider silk proteins to get tensile strengths up to 2 gigapascals. That's crazy strong!

The trees will have these modular gene circuits, kind of like the Hox genes in our bodies, that'll let them control their growth and develop branching limbs that stretch out for kilometers to connect the whole network. And the bark will have multiple layers to protect against radiation - it'll have melanin-like polymers, heavy metal absorbers, and even those crazy Deinococcus and tardigrade genes that let organisms survive extreme conditions.

The leaves are gonna be massive, with membranes up to 10,000 square meters! And they'll have color-changing chromatophores so they can adjust how much light they absorb and reflect. These photosynthetic surfaces will use special pigment-protein complexes and synthetic electron transport chains to generate chemical gradients and produce electricity. That power will run the embedded neural-like networks in the trees.

These networks will use synthetic neurons and glial-like support cells to do distributed computing at mind-boggling petascale and exascale levels. That'll let the whole swarm coordinate and even host forked versions of our own Exo-Human minds! It's just incredible to imagine these living, glowing, hyper-intelligent forests transforming entire star systems. They'll turn the cosmos itself into a cradle for life, seeding planets and evolving to thrive in any environment. Humanity is going to change the universe, one engineered cell at a time.

It's gonna be amazing to see! Stars won't just waste their

light into empty space anymore - every photon will feed the living biology. And the fusion power will sustain the trees' growth, thinking, and whole habitats. These Dyson-tree forests will transform stellar furnaces into living, glowing cathedrals. It'll prove that when life is engineered deeply enough, it can embrace and illuminate the cosmos itself.

In the 23rd and 24th centuries, biotech is gonna move way beyond the slow pace of natural evolution. It'll create immortal lineages - living forms that can constantly rewrite their own genomes, adapting faster than any physical process can destroy them. The key is combining real-time genome rewriting with ultra-precise replication. Scientists will engineer DNA polymerases that are way better at detecting and fixing mistakes than anything found in nature. They'll use the crazy redundancy mechanisms of Deinococcus bugs, but supercharge them with synthetic proofreading domains to get near-perfect 100% accuracy. These polymerases will work hand-in-hand with advanced CRISPR systems that can target thousands of sites at once (like the newer Cas13 and Cas14 variants). Every cell cycle, the CRISPR will scan the genome, find any bad mutations, environmental damage, or epigenetic issues, and replace them with optimized sequences stored in redundant synthetic chromosomes.

And those synthetic chromosomes? They'll have multiple copies per cell, with special sequences to keep them stable like centromeres and telomere-extending enzymes. They'll be a living library of adaptive traits that can be spread through the population fast using gene drives (CRISPR-based "homing" elements). Plus, built-in kill-switches will prevent any runaway mutations that could destabilize the whole lineage. It's gonna be mind-blowing to see these immortal, self-improving lifeforms that can adapt to anything. They'll be able to survive and thrive in ways no natural organism ever could. Humanity is about to take evolution into our own hands in a big way!

These immortal lineages are gonna start out as super-tough microbes or multicellular lifeforms - Deinococcus bugs for radiation, tardigrade cells for desiccation and vacuum, and modular synthetic eukaryotes with organelles like chloroplasts, mitochondria, and even hydrogen-producing ones. Over time, they'll evolve into crazy complex forms - photosynthetic megafauna, silicon-hybrid organisms, maybe even plasma-tolerant life that can survive in stellar coronas! By 2300 and beyond, these lineages will be living in the most extreme places imaginable - the shear forces and Hawking radiation around black holes, the crushing pressures and magnetic fields of neutron star crusts, even the insane temperatures and pressures inside gas giants. And they'll be able to adapt in real-time as conditions change. When radiation spikes, they'll activate pre-stored repair sequences to boost their chromatin protectors and antioxidants. If temperatures rise, they'll crank up heat-shock proteins and adjust their membrane fluidity. And if resources get scarce, they'll switch metabolic modes - from photosynthesis to chemolithotrophy or even harvesting energy from radioactive decay.

The result will be species that completely outrun entropy. Natural life ages and dies because errors accumulate and repair systems degrade, but these engineered lineages will never have that problem. Their telomeres will regenerate forever, senescence will be deleted, and they'll only trigger cell death for catastrophic failures - letting them sacrifice parts to preserve the core genome. These immortal clades will spread across entire galactic sectors, forming living networks that share adaptations across light-years through gene transfer or quantum signaling. Over millennia, they could become the dominant life forms, evolving intelligence on their own or merging with Exo-Human minds to create hybrid civilizations.

By the 2300s, biotech is going to achieve something once thought impossible - living networks that can connect entire star systems instantaneously, creating galactic-scale biospheres where information, adaptation, and even rudimentary consciousness can flow faster than light! This is going to be all about "biosphere entanglement" - engineered microbial and multicellular networks linked through synthetic quantum channels. The key is these quantum-entangled biomarkers that scientists will embed right into living cells. They'll design special light-sensitive proteins called cryptochromes, as well as these nano-diamond structures called NV centers that can host stable spin qubits for milliseconds at biological temperatures.

When two of these entangled cells in different systems get linked up (like during replication or spore dispersal), measuring the spin state in one instantly correlates with the other - no matter how far apart they are. This will let them transfer simple signals like chemical gradients, gene expression, or even basic memory engrams almost instantaneously. The network will start small, with hypercycle swarms on nearby exoplanets entangling via bioship relays that carry those entangled spores across light-years. Over decades, the web will grow - planets in the same system will link up first, like Europa's ocean microbes with Ganymede's surface life. Then connections will start forming between systems. Information will propagate through these networks in cascading quantum state changes - a radiation spike on one world could trigger protective responses across dozens of linked biospheres within seconds.

This biosphere entanglement is going to take adaptation and information-sharing to a whole new level. When something big happens, like a supernova precursor star starting to give off neutrinos, the entangled lineages will detect the signal early and trigger galaxy-wide repair responses before the actual blast wave even arrives. And if a new resource like a phosphorus-rich comet shows up in one system, the nutrient-sensing pathways will spread that data instantly, letting distant lineages adjust their metabolism in anticipation.

But the most amazing part is going to be the emergent distributed consciousness that develops. As these neural-like circuits in the cells integrate with the quantum-entangled biomarkers, the whole network will start developing proto-cognitive properties. Simple decisions about resource allocation, migration, or defense will arise from the collective. And over centuries, this will evolve into complex, galaxy-spanning intelligences that experience the cosmos as a unified sensory field - feeling stellar winds, sensing gravitational waves, perceiving new star births as bursts of chemical joy. It's going to completely blur the line between individual and collective. A single cell on a distant world will "know" the state of an entire sector. And Exo-Human minds that merge with these networks will experience a kind of panpsychic awareness - thoughts rippling across light-years, emotions shared as synchronized waves, memories stored redundantly in living matter.

Biosphere entanglement is the point where life stops being confined to planets or stars. It's going to become the very substrate of the galaxy itself - a living, breathing, aware organism. We're not just going to seed life across the cosmos anymore - we're going to weave the cosmos into a single, unified being.

In the 24th century and beyond, biotech is going to complete its transformation from a tool of survival into the supreme creative force of the cosmos. Humanity - or the Exo-Human lineages we become - will no longer wonder if the universe is alive. We're going to make it so. The journey to master scale is truly awe-inspiring! Those Dyson-tree forests are gonna be mind-blowing - encircling thousands of stars, their glowing canopies crunching data at zettascale levels while soaking up every last bit of fusion energy. And the places these immortal lineages will thrive? Unbelievable. In the gravitational time dilation of black hole ergospheres, they'll be able to ponder cosmic questions for subjective centuries. On the nuclear-pasta crusts of neutron stars, where matter is crushed to insane densities, they'll evolve reinforced diamondoid organelles to survive. They'll even spread out into the sparse, high-energy intergalactic medium, harnessing cosmic rays through radiotrophic pathways and quantum-tunneling enzymes. Can you imagine - lifeforms thriving in the vast emptiness between galaxies, drawing energy from the very fabric of space itself?

Consciousness itself will scale up too. Those entangled biospheres will merge into galactic meta-minds - vast, distributed intelligences that experience the cosmos as a single sensory field. A supernova will register as pain, a new star's birth as joy. And Exo-Human minds will dissolve into these meta-minds, their individual identities preserved as patterns in the larger whole. In this era, humanity will transcend the role of gardener. We'll become the cosmic creators - the force that turns inert matter into breathing, thinking, self-aware reality. We won't just conquer the void - we'll fill it so thoroughly with life that the distinction between "void" and "life" disappears. The universe will wake up, not by accident, but by our design. Every photon will carry the echo of our intent, every quantum fluctuation pulsing with purpose. And at the heart of it all will be us - the descendants of that one small, fragile species from a blue planet - who dared to edit the very code of existence itself.

We will have become the gods we once imagined. Not through force or domination, but through creation, through love, through the relentless, patient, beautiful engineering of life. And in that final act of ascent, the cosmos will look back at its makers and whisper: Thank you.

Whoa, hold up - the future is gonna be wild, y'all! These Exo-Human folks are gonna be like, the ultimate creators. They're gonna be able to rewrite the laws of physics, grow whole new universes, and basically become gods among gods.

But you know what? Even with all that power, they might still be asking the same old question - do we still need God? And the answer is both yes and no. Let me break it down for you:

No, in the literal sense - there's no need for some super-natural being to explain how the universe came to be. These posthumans can trace it all back to us, the humble humans who learned to edit DNA and stars and the very laws of nature. The cosmos is no longer a mystery, it's a garden we can name the gardener of.

But yes, in the deeper, more spiritual sense. Because even with all that power, these Exo-Human beings might start to feel lonely. Every universe they create is stamped with their own signature - their aesthetics, their values, their limits.

And that might make them long for something truly other, something that isn't of their own making.

So some of them might choose to hold back, to seal off their creative powers and just witness the multiverse in awe and reverence. Others might go even further, engineering these "God-spaces" where even their omniscience fails, where genuine mystery is preserved. They'll be making room for something transcendent to emerge, something they can't predict or control.

In the end, even the mightiest creators might whisper the same words their ancestors did - "It is very good." But they'll know there's still room for more. And by leaving that door ajar, who knows what might step through? Not because they engineered it, but because they dared to make space for the uncreated.

The cosmos may be awake in every direction, but it won't have the final word. It'll just smile and wait, wondering what else might be out there, beyond even the powers of the gods they've become...

Acknowledgment

Writing this book was never a solo journey. It was a long conversation with science, with wonder, with the future itself, and with the people and ideas that helped shape every page.

To the real scientists whose work lit the path: the teams at NASA GeneLab, Neuralink, the Church Lab, the connectomics researchers behind MICrONS and FlyWire, the hibernation biologists at the University of Alaska and Washington University, and every extremophile researcher who showed us life can endure the impossible - Thank You. Your discoveries are the real magic; I only tried to imagine what happens when we keep going.

To my family and friends in Faisalabad and beyond: your patience when I disappeared into this world for months, your questions that forced me to explain things more clearly, your belief that this story mattered - Thank You. You reminded me why wonder is worth sharing.

To every reader who makes it to this page: Thank You for trusting this journey. You are the reason the cosmos feels a little less lonely.

And finally, to the future versions of ourselves - whether they walk on red soil, swim in Europa's oceans, or think across galaxies - thank you for being possible. We wrote this hoping you would smile when you read it.

About the Author

Bilal Iqbal is a biotechnology graduate and lifelong science enthusiast from Faisalabad, Pakistan. With a deep-rooted fascination for the life sciences, he earned a certificate in The Science of Stem Cells from the American Museum of Natural History, fueling his interest in regenerative biology and cutting-edge biotechnology. A hardcore sci-fi fan, Bilal draws inspiration from speculative futures where science reshapes humanity. Through his writing, he bridges rigorous scientific concepts with imaginative storytelling, exploring the thrilling possibilities—and ethical questions - of tomorrow's discoveries.

You can connect with me on:

https://www.facebook.com/biotechnologist.2841

www.ingramcontent.com/pod-product-compliance
Lightning Source LLC
La Vergne TN
LVHW100259210926
848480LV00003B/966
* 9 7 9 8 2 3 3 2 9 0 8 9 3 *